Algebra 1

LARSON
BOSWELL
KANOLD
STIFF

Applications • Equations • Graphs

Chapter 5 Resource Book

The Resource Book contains the wide variety of blackline masters available for Chapter 5. The blacklines are organized by lesson. Included are support materials for the teacher as well as practice, activities, applications, and assessment resources.

Contributing Authors

The authors wish to thank the following individuals for their contributions to the Chapter 5 Resource Book.

Rita Browning
Linda E. Byrom
José Castro
Christine A. Hoover
Carolyn Huzinec
Karen Ostaffe
Jessica Pflueger
Barbara L. Power
Joanne Ricci
James G. Rutkowski
Michelle Strager

ISBN: 0-618-02043-8

15 14 13 12 11 10 -CKI- 06 05 04

Contents

Contents

Contents

Descriptions of Resources

This Chapter Resource Book is organized by lessons within the chapter in order to make your planning easier. The following materials are provided:

Tips for New Teachers These teaching notes provide both new and experienced teachers with useful teaching tips for each lesson, including tips about common errors and inclusion.

Parent Guide for Student Success This guide helps parents contribute to student success by providing an overview of the chapter along with questions and activities for parents and students to work on together.

Prerequisite Skills Review Worked-out examples are provided to review the prerequisite skills highlighted on the Study Guide page at the beginning of the chapter. Additional practice is included with each worked-out example.

Strategies for Reading Mathematics The first page teaches reading strategies to be applied to the current chapter and to later chapters. The second page is a visual glossary of key vocabulary.

Lesson Plans and Lesson Plans for Block Scheduling This planning template helps teachers select the materials they will use to teach each lesson from among the variety of materials available for the lesson. The block-scheduling version provides additional information about pacing.

Warm-Up Exercises and Daily Homework Quiz The warm-ups cover prerequisite skills that help prepare students for a given lesson. The quiz assesses students on the content of the previous lesson. (Transparencies also available)

Activity Support Masters These blackline masters make it easier for students to record their work on selected activities in the Student Edition.

Alternative Lesson Openers An engaging alternative for starting each lesson is provided from among these four types: ***Application, Activity, Graphing Calculator,*** or ***Visual Approach.*** (Color transparencies also available)

Graphing Calculator Activities with Keystrokes Keystrokes for four models of calculators are provided for each Technology Activity in the Student Edition, along with alternative Graphing Calculator Activities to begin selected lessons.

Practice A, B, and C These exercises offer additional practice for the material in each lesson, including application problems. There are three levels of practice for each lesson: A (basic), B (average), and C (advanced).

Contents

Reteaching with Practice These two pages provide additional instruction, worked-out examples, and practice exercises covering the key concepts and vocabulary in each lesson.

Quick Catch-Up for Absent Students This handy form makes it easy for teachers to let students who have been absent know what to do for homework and which activities or examples were covered in class.

Cooperative Learning Activities These enrichment activities apply the math taught in the lesson in an interesting way that lends itself to group work.

Interdisciplinary Applications/Real-Life Applications Students apply the mathematics covered in each lesson to solve an interesting interdisciplinary or real-life problem.

Math and History Applications This worksheet expands upon the Math and History feature in the Student Edition.

Challenge: Skills and Applications Teachers can use these exercises to enrich or extend each lesson.

Quizzes The quizzes can be used to assess student progress on two or three lessons.

Chapter Review Games and Activities This worksheet offers fun practice at the end of the chapter and provides an alternative way to review the chapter content in preparation for the Chapter Test.

Chapter Tests A, B, and C These are tests that cover the most important skills taught in the chapter. There are three levels of test: A (basic), B (average), and C (advanced).

SAT/ACT Chapter Test This test also covers the most important skills taught in the chapter, but questions are in multiple-choice and quantitative-comparison format. (See *Alternative Assessment* for multi-step problems.)

Alternative Assessment with Rubrics and Math Journal A journal exercise has students write about the mathematics in the chapter. A multi-step problem has students apply a variety of skills from the chapter and explain their reasoning. Solutions and a 4-point rubric are included.

Project with Rubric The project allows students to delve more deeply into a problem that applies the mathematics of the chapter. Teacher's notes and a 4-point rubric are included.

Cumulative Review These practice pages help students maintain skills from the current chapter and preceding chapters.

CHAPTER 5

Tips for New Teachers

For use with Chapter 5

Lesson 5.1

COMMON ERROR Students might try to graph problems such as Example 3 on page 274 by using the slope and y-intercept of the line, as they learned in Lesson 4.6, Quick Graphs Using Slope-Intercept Form. Remind them that this technique can only be used if the scales used for the horizontal and vertical axes are the same. Otherwise they must graph the line by plotting two points.

TEACHING TIP The examples for real-life situations in the lesson can be solved without making a graph. Ask your students to name some advantages of making one. For example, using the graph it is easy to tell whether the linear model is increasing or decreasing. In addition, the graph can be used to quickly estimate other values of the linear function. You can use this discussion when introducing Lesson 5.7, Predicting with Linear Models.

TEACHING TIP Show your students that saying that b is the initial amount is equivalent to saying that b is the value of the function when x, the independent variable, is zero. This fact might influence what we choose for the independent variable, because the point $(0, b)$ must be on the graph of the line (see Example 3 on page 274).

Lesson 5.2

TEACHING TIP Start the lesson by asking your students whether the slope of a line and a point on the line are enough information to graph it. Since this is possible, it should also be possible to find an equation for such a line if the same information is known. Lesson 5.2 will show your students how to find that equation. This approach emphasizes the connection between the equation and the graph of a linear function.

TEACHING TIP Use real-life situations such as Example 3 on page 281 to discuss again domain and range of functions. These problems usually have restrictions in either the domain or the range. Use these examples to have students check whether the answers they obtain make sense. For example, would a negative number of trips make sense?

Lesson 5.3

TEACHING TIP Ask your students if two points are enough information to graph a line. Since the answer is yes, it must also be possible to find the equation of a line given two points on it.

TEACHING TIP After finding the slope of the line, some students wonder which point they should use to find the y-intercept b. To show them that it does not matter, give two points on a line. Have one half of your class find the equation of the line by using the first point to find b and have the other half use the second point. They should all get the same answer.

TEACHING TIP Include some class examples where the two given points lay on a horizontal or a vertical line. Once the students find the corresponding slope, whether zero or undefined, they should realize that these are special lines. Review how to write and graph the equations for vertical and horizontal lines.

COMMON ERROR You might want to use "opposite reciprocals" instead of "negative reciprocals" to describe the slopes of perpendicular lines. Otherwise, some students might think that the slopes are always negative for both lines.

Lesson 5.4

COMMON ERROR Many students believe that the best-fitting line goes through the leftmost and rightmost points of their graph, ignoring all the other points. Point out that a "good" best-fitting line should go through as many data points as possible. If it cannot go through all points, it should leave above and below the line an approximately equal number of data points. This means that some data points might not be on the line, including the leftmost and the rightmost.

COMMON ERROR Some students believe that they must first choose two data points from their graph before they draw the best-fitting line. Remind students that they must first draw the best-fitting line and then choose two points on it to find its equation. Also, remind them that these two points do not have to be data points.

CHAPTER 5 CONTINUED

Tips for New Teachers

For use with Chapter 5

TEACHING TIP Draw some scatter plots where the data points can be approximated by either a vertical or a horizontal line. Discuss with your students whether there is correlation and, if so, of what type. Ask them for real examples that would result in those types of scatter plots.

Lesson 5.5

COMMON ERROR Some students may have trouble working with the point-slope form of a linear equation. They might plug in values for x instead of x_1 or for y instead of y_1. Tell your students that a letter with a subscript, such as x_1 or y_1, represents a specific point and its coordinate values must be substituted into the equation. The final equation should contain the variables x and y.

COMMON ERROR Another problem students might have with the point-slope form of the linear equation is that they might forget to account for the minus signs in the formula or reverse the x and y coordinates as (y_1, x_1).

Lesson 5.6

COMMON ERROR Some students believe that the different forms of a linear equation represent different lines. The Activity for Developing Concepts on page 308 will help these students to understand that the same line can be written in several different forms. The forms of a linear equation can be compared to the representations of a number. For instance, the number five can be represented by spelling it out as "five," by the numeral "5," by the roman numeral "V," or even by drawing five dots as in a die. These are all just different representations of the same thing.

TEACHING TIP Ask students what technique they would use to graph a linear equation based on the form in which it was given to them. Show them how finding the intercepts is an easy method when the equation is in standard form. This will help to review the graphing methods covered in Chapter 4 and will show students that they do not always have to find the slope-intercept form of a line to graph it.

Lesson 5.7

TEACHING TIP Sometimes students can say whether a linear model can be used for the data in a table without making a scatter plot. All they must check is that equal interval values of the independent variable result in equal interval values of the dependent variable. For instance, in Example 1 on page 316, the years always increase by 2. The data for broadcasting television increase by approximately equal intervals of about 4500 million. However, the data for the Internet do not show equal increasing intervals. Therefore, the broadcasting television data fit a linear model better.

INCLUSION You can help students with limited English proficiency to learn new vocabulary words by making connections to words they already know. For instance, interpolation is related to internal, which means inside. Similarly, extrapolation is related to external, which means outside. This might help students to remember the difference between interpolation and extrapolation.

Outside Resources

BOOKS/PERIODICALS

VanDyke, Frances. "Relating to Graphs in Introductory Algebra." *Mathematics Teacher* (September 1994); pp. 427–432, 438, 439.

ACTIVITIES/MANIPULATIVES

Anderson, Edwin D. and Jim Nelson. "An Introduction to the Concept of Slope." *Mathematics Teacher* (January 1994); pp. 27–30, 37–41.

SOFTWARE

Dugdale, Sharon and David Kibbey. *Green Globs and Graphing Equations*. Introductory graphing concepts, tutorials, exploring graphs. Pleasantville, NY; Sunburst Communications.

VIDEOS

Algebra in Simplest Terms. Linear equations. Burlington, VT; Annenburg/CPB Collection, 1991.

NAME ______________________________ DATE ____________

Parent Guide for Student Success

For use with Chapter 5

Chapter Overview One way that you can help your student succeed in Chapter 5 is by discussing the lesson goals in the chart below. When a lesson is completed, ask your student to interpret the lesson goals for you and to explain how the mathematics of the lesson relates to one of the key applications listed in the chart.

Lesson Title	Lesson Goals	Key Applications
5.1: Writing Linear Equations in Slope-Intercept Form	Use the slope-intercept form to write the equation of a line. Model a real-life situation with a linear function.	• Population Change • Renting a Moving Van • Fundraising
5.2: Writing Linear Equations Given the Slope and a Point	Use the slope and any point on a line to write an equation of the line. Use a linear model to make predictions about a real-life situation.	• Vacation Trips • Cellular Rates • Taxi Ride
5.3: Writing Linear Equations Given Two Points	Write an equation of a line given two points on the line. Use a linear equation to model a real-life problem.	• Archaeology • Chunnel • Echoes
5.4: Fitting a Line to Data	Find a linear equation that approximates a set of data points. Determine the type of correlation in a set of real-life data.	• Biology • Discus Throws • Football Salaries
5.5: Point-Slope Form of a Linear Equation	Use the point-slope form to write an equation of a line and to model a real-life situation.	• Marathon • Field Trip • Mountain Climbing
5.6: The Standard Form of a Linear Equation	Write a linear equation in standard form and use the standard form to model real-life situations.	• Barbecue • Bird Seed Mixture • Color Printing
5.7: Predicting with Linear Models	Determine whether a linear model is appropriate. Use a linear model to make a real-life prediction.	• Advertising • Movie Theaters • College Tuition

Study Strategy

Writing and Taking a Practice Test is the study strategy featured in Chapter 5 (see page 272). Encourage your student to create a test, exchange tests with a classmate, score each other's tests, and plan further study. You may wish to discuss with your student the kinds of test questions the teacher is likely to ask and to help your student in developing a study plan.

CHAPTER 5 CONTINUED

NAME ______________________________ DATE ____________

Parent Guide for Student Success

For use with Chapter 5

Key Ideas Your student can demonstrate understanding of key concepts by working through the following exercises with you.

Lesson	*Exercise*
5.1	A cellular phone costs \$15 a month plus \$0.18 per minute of use. Write a linear equation to model the cost of cellular phone use and find the monthly bill for 20 minutes of use.
5.2	Write an equation of the line that is parallel to $y = \frac{1}{2}x - 9$ and passes through the point $(-4, 7)$.
5.3	Write an equation in slope-intercept form of the line that passes through the points $(-2, 9)$ and $(6, -7)$.
5.4	Would you expect a *positive correlation*, a *negative correlation*, or *relatively no correlation* between the value of a dollar and the years since 1900? Explain.
5.5	Write an equation in point-slope form of the line that passes through the points $(-5, -4)$ and $(-2, 8)$.
5.6	You have \$60 to spend at a sale where sweatshirts are \$10 each and blue jeans are \$15 each. Your parents agree to pay the tax. Write an equation that models the different number of sweatshirts and blue jeans you can buy. Find at least three combinations that allow you to spend the whole \$60.
5.7	In 1970, the leading jockey won about \$2.63 million. In 1980, the leader won about \$7.66 million, and in 1990, the leader won about \$13.88 million. Write a linear model for the amount earned y by the leading jockey t years after 1970. Use it to predict the amount earned in the year 2000.

Home Involvement Activity

You will need: A calculator

Directions: Find the amount your family spent on utilities (or one specific utility) last year and at least 3 years ago. Use the data to find a linear model for the amount y spent t years after 1990. Use the model to estimate how much you can expect to spend on utilities next year.

Answers

5.1: $y = 0.18t + 15$; \$18.60 **5.2:** $y = \frac{1}{2}x + 9$ **5.3:** $y = -2x + 5$ **5.4:** negative correlation; the value of a dollar has been decreasing as the years since 1900 have increased. **5.5:** $y + 4 = 4(x + 5)$ or $y - 8 = 4(x + 2)$ **5.6:** $10x + 15y = 60$; 6 sweatshirts and no jeans, no sweatshirts and 4 jeans, 3 sweatshirts and 2 jeans **5.7:** *Sample answer:* $y = 0.563x + 2.43$; about \$19.3 million

CHAPTER 5

NAME ______________________ DATE __________

Prerequisite Skills Review

For use before Chapter 5

EXAMPLE 1 Solving Equations With Variables on Both Sides

Solve the equation.

$8(x - 1) = 4x - 17 + x$

SOLUTION

$8(x - 1) = 4x - 17 + x$	Write the original equation.
$8x - 8 = 4x - 17 + x$	Use the distributive property.
$8x - 8 = 5x - 17$	Add like terms.
$3x - 8 = -17$	Subtract $5x$ from each side.
$3x = -9$	Add 8 to each side.
$x = -3$	Divide each side by 3.

Exercises for Example 1

Solve the equation.

1. $2(1 - x) + 3x = -4(x + 2)$

2. $12x - 8 = 3(4x + 11)$

3. $\frac{1}{5}(25x + 60) = 33 - 4(x - 6)$

4. $-\frac{2}{9}(18x - 9) = 6\left(x - \frac{1}{2}\right)$

EXAMPLE 2 Plotting Points in a Coordinate Plane

Plot the ordered pairs in a coordinate plane.

$A(6, -1)$, $B(-3, -7)$, $C(0, 2)$

SOLUTION

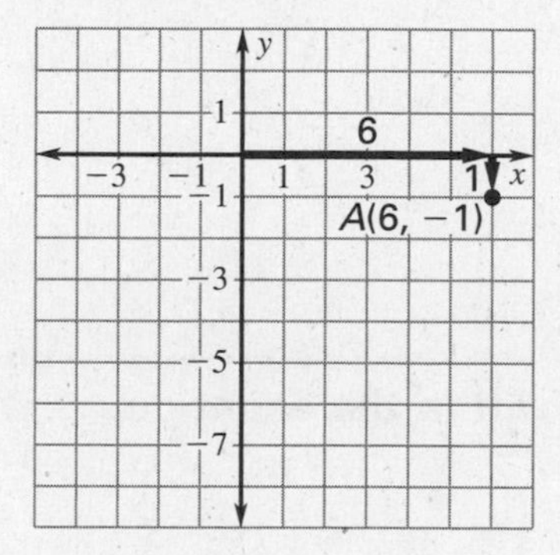

To plot the point $(6, -1)$ start at the origin. Move 6 units to the right and 1 unit down.

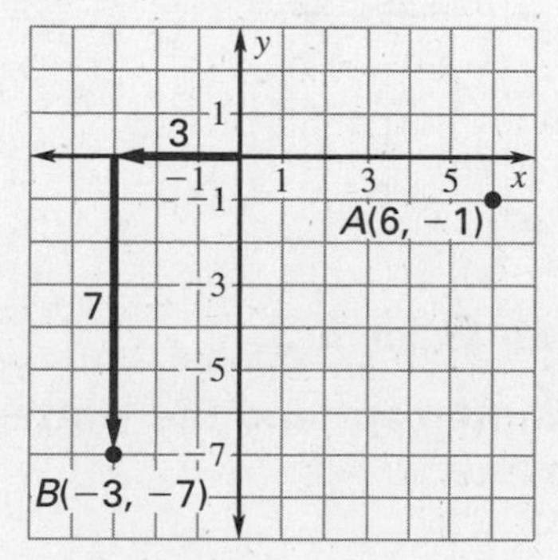

To plot the point $(-3, -7)$ start at the origin. Move 3 units to the left and 7 units down.

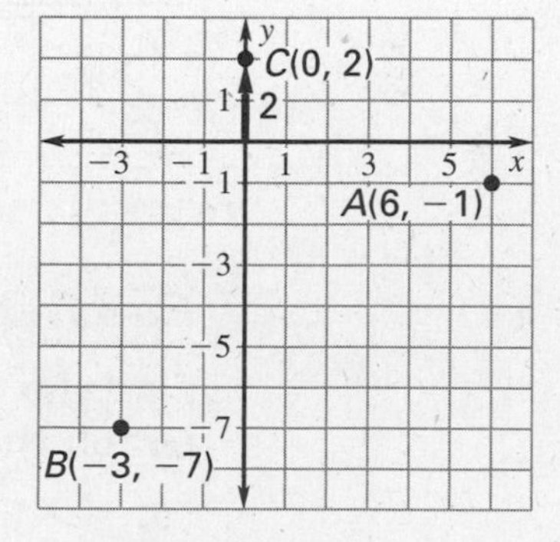

To plot the point $(0, 2)$ start at the origin. Move 2 units up.

NAME ______________________________ DATE ____________

Prerequisite Skills Review

For use before Chapter 5

Exercises for Example 2

Plot the ordered pairs in a coordinate plane.

5. $A(5, 2), B(-3, 8), C(7, -2)$

6. $A(-4.1, -3), B(-1, -1), C(2.1, 4)$

7. $A(-1, 0), B(0, -6), C\left(\frac{1}{2}, -4\right)$

8. $A(-3, 0), B(1.5, 3), C(5, -3.5)$

EXAMPLE 3 Using Intercepts to Graph Equations

Find the x-intercept and the y-intercept of the equation. Graph the line.

$5x + 8y = 24$

SOLUTION

Find the intercepts by substituting 0 for y and then 0 for x.

$$\begin{aligned} 5x + 8y &= 24 \\ 5x + 8(0) &= 24 \\ 5x &= 24 \\ x &= \tfrac{24}{5} \end{aligned}$$

$$\begin{aligned} 5x + 8y &= 24 \\ 5(0) + 8y &= 24 \\ 8y &= 24 \\ y &= 3 \end{aligned}$$

The x-intercept is $\frac{24}{5} = 4\frac{4}{5}$. The y-intercept is 3.

Draw a line that includes the points $\left(\frac{24}{5}, 0\right)$ and $(0, 3)$.

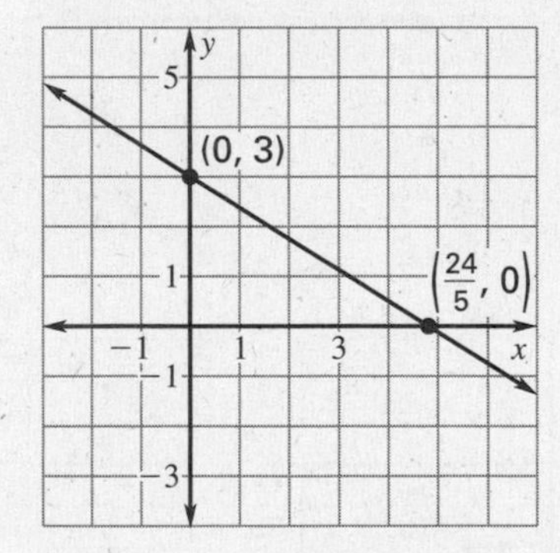

Exercises for Example 3

Find the x-intercept and the y-intercept in the equation. Graph the line.

9. $5x + y = 25$

10. $-4x = 1.1y - 3.3$

11. $-x - 6y = 35 + 4x$

12. $x + 8y = 10$

NAME ______________________________ DATE ____________

Strategies for Reading Mathematics

For use with Chapter 5

Strategy: Translating Words into Symbols

When you use mathematics to solve real-life problems, you will often translate the words into symbols. Read the following problem.

> A family pays a one-time registration fee of $35 to enroll their child in daycare. The weekly charge thereafter is $200 per week. How much will the family pay for daycare?

The steps outlined below can help you interpret this problem and other problems like it.

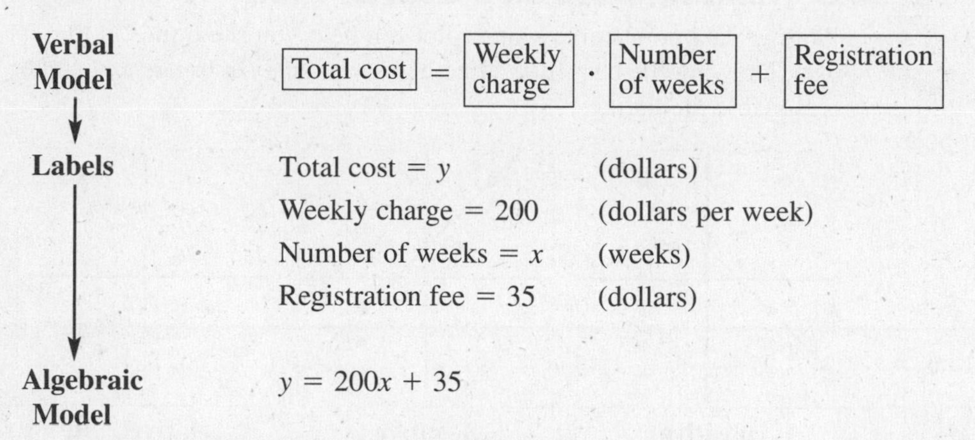

STUDY TIP

Use a Problem-Solving Model

To model a real-life situation using mathematics, begin by writing a verbal model based on the relationships in the situation. Next assign labels to the pieces of your model. Then use your labels to write an algebraic model for the situation.

Questions

1. How much does it cost for three weeks of daycare? for five weeks of daycare? What is the rate of change in dollars per week?
2. If you graphed the equation, what would be the y-intercept?
3. Suppose the registration fee was $40 and the weekly charge was $225. Write an equation to represent this situation.
4. Write a verbal model, labels, and an algebraic model for the following situation.

   Rachel receives a weekly allowance of $3.00 for doing certain household chores. She can earn more money by doing additional chores for $1.50 per chore. How much money can Rachel earn in a week?

NAME ______________________ DATE ____________

Strategies for Reading Mathematics

For use with Chapter 5

Visual Glossary

The Study Guide on page 272 lists the key vocabulary for Chapter 5 as well as review vocabulary from previous chapters. Use the page references on page 272 or the Glossary in the textbook to review key terms from prior chapters. Use the visual glossary below to help you understand some of the key vocabulary in Chapter 5. You may want to copy these diagrams into your notebook and refer to them as you complete the chapter.

GLOSSARY

best-fitting line (p. 292) A line that best fits the data points on a scatter plot.

correlation (p. 295) The relationship between two data sets.

slope-intercept form (pp. 241, 273) A linear equation written in the form $y = mx + b$. The slope of the line is m. The y-intercept is b.

point-slope form (p. 300) The equation of a nonvertical line $y - y_1 = m(x - x_1)$ that passes through a given point (x_1, y_1) with a slope of m.

standard form of an equation of a line (p. 308) A linear equation of the form $Ax + By = C$ where A, B, and C are integers and A and B are not both zero.

Correlating Data on a Scatter Plot

Plotting data points on a scatter plot can help you see trends in the data. Drawing the best-fitting line helps you analyze trends and make reasonable predictions.

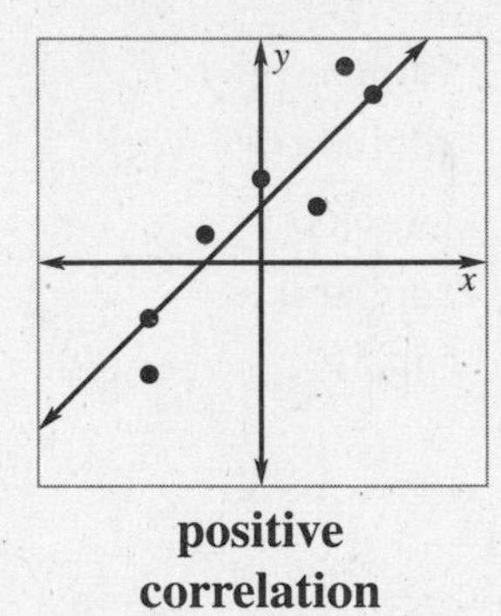

positive correlation

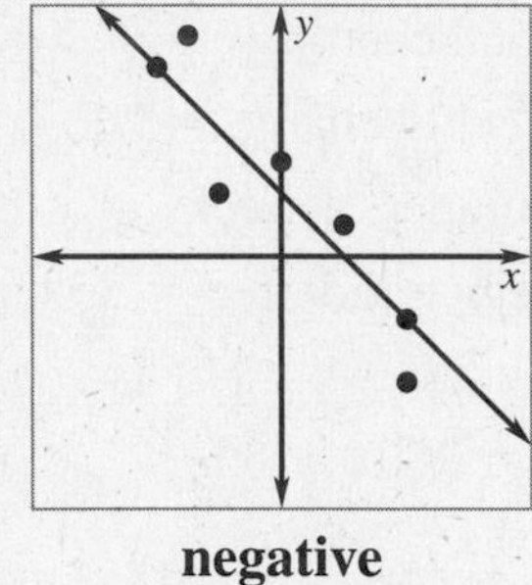

negative correlation

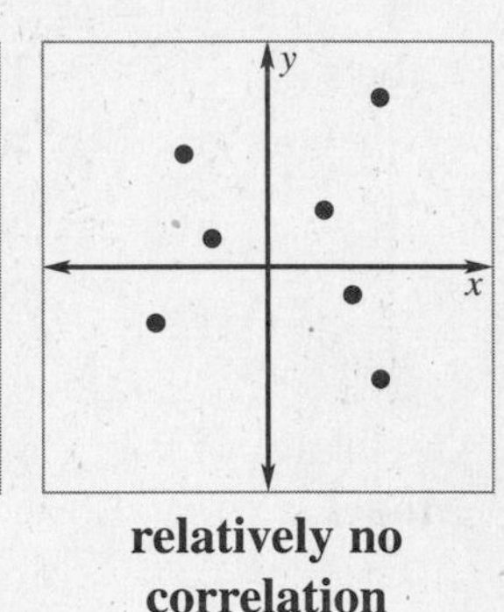

relatively no correlation

Graphing Equations of a Line

You can use a graph or an equation of a best-fitting line to make predictions. There are three common forms of a linear equation.

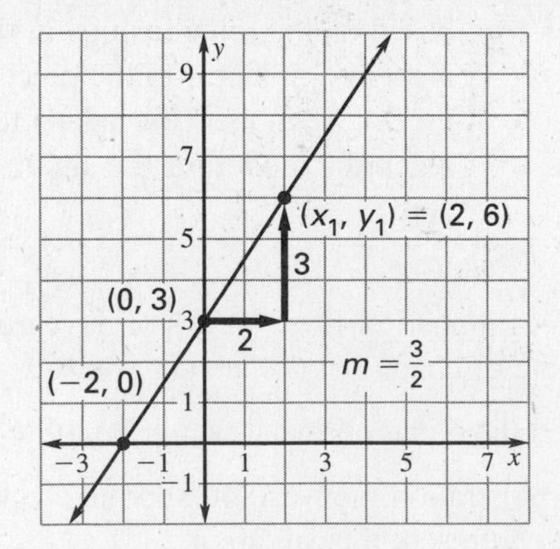

slope-intercept form	**point-slope form**	**standard form**
$y = mx + b$	$y - y_1 = m(x - x_1)$	$Ax + By = C$
$y = \frac{3}{2}x + 3$	$y - 6 = \frac{3}{2}(x - 2)$	$-3x + 2y = 6$

LESSON

TEACHER'S NAME ______________________ CLASS ______ ROOM ______ DATE ______

Lesson Plan

1-day lesson (See *Pacing the Chapter,* TE pages 270C–270D) **For use with pages 273–278**

GOALS
1. Use the slope-intercept form to write an equation of a line.
2. Model a real-life situation with a linear function.

State/Local Objectives ____

____

✓ Check the items you wish to use for this lesson.

STARTING OPTIONS

____ Prerequisite Skills Review: CRB pages 5–6
____ Strategies for Reading Mathematics: CRB pages 7–8
____ Warm-Up or Daily Homework Quiz: TE pages 273 and 262, CRB page 11, or Transparencies

TEACHING OPTIONS

____ Lesson Opener (Graphing Calculator): CRB page 12 or Transparencies
____ Examples 1–4: SE pages 273–275
____ Extra Examples: TE pages 274–275 or Transparencies
____ Closure Question: TE page 275
____ Guided Practice Exercises: SE page 276

APPLY/HOMEWORK

Homework Assignment

____ Basic 12–24 even, 26–32, 38, 39, 42, 44–51
____ Average 12–24 even, 26–33, 38, 39, 42, 44–51
____ Advanced 12–24 even, 30–40, 42, 44–51

Reteaching the Lesson

____ Practice Masters: CRB pages 13–15 (Level A, Level B, Level C)
____ Reteaching with Practice: CRB pages 16–17 or Practice Workbook with Examples
____ Personal Student Tutor

Extending the Lesson

____ Applications (Interdisciplinary): CRB page 19
____ Challenge: SE page 278; CRB page 20 or Internet

ASSESSMENT OPTIONS

____ Checkpoint Exercises: TE pages 274–275 or Transparencies
____ Daily Homework Quiz (5.1): TE page 278, CRB page 23, or Transparencies
____ Standardized Test Practice: SE page 278; TE page 278; STP Workbook; Transparencies

Notes ____

____

____

TEACHER'S NAME ____________ CLASS ______ ROOM ______ DATE ______

Lesson Plan for Block Scheduling

Half-day lesson (See *Pacing the Chapter,* TE pages 270C–270D) **For use with pages 273–278**

GOALS
1. **Use the slope-intercept form to write an equation of a line.**
2. **Model a real-life situation with a linear function.**

State/Local Objectives ____________

CHAPTER PACING GUIDE	
Day	**Lesson**
1	**5.1 (all)**; 5.2 (all)
2	5.3 (all)
3	5.4 (all); 5.5 (begin)
4	5.5 (end); 5.6 (begin)
5	5.6 (end); 5.7 (begin)
6	5.7 (end); Review Ch. 5
7	Assess Ch. 5; 6.1 (all)

✓ **Check the items you wish to use for this lesson.**

STARTING OPTIONS

____ Prerequisite Skills Review: CRB pages 5–6
____ Strategies for Reading Mathematics: CRB pages 7–8
____ Warm-Up or Daily Homework Quiz: TE pages 273 and 262, CRB page 11, or Transparencies

TEACHING OPTIONS

____ Lesson Opener (Graphing Calculator): CRB page 12 or Transparencies
____ Examples 1–4: SE pages 273–275
____ Extra Examples: TE pages 274–275 or Transparencies
____ Closure Question: TE page 275
____ Guided Practice Exercises: SE page 276

APPLY/HOMEWORK

Homework Assignment (See also the assignment for Lesson 5.2.)

____ Block Schedule: 12–24 even, 26–33, 38, 39, 42, 44–51

Reteaching the Lesson

____ Practice Masters: CRB pages 13–15 (Level A, Level B, Level C)
____ Reteaching with Practice: CRB pages 16–17 or Practice Workbook with Examples
____ Personal Student Tutor

Extending the Lesson

____ Applications (Interdisciplinary): CRB page 19
____ Challenge: SE page 278; CRB page 20 or Internet

ASSESSMENT OPTIONS

____ Checkpoint Exercises: TE pages 274–275 or Transparencies
____ Daily Homework Quiz (5.1): TE page 278, CRB page 23, or Transparencies
____ Standardized Test Practice: SE page 278; TE page 278; STP Workbook; Transparencies

Notes ____________

LESSON **5.1**

NAME ________________ DATE ________

Available as a transparency

WARM-UP EXERCISES

For use before Lesson 5.1, pages 273–278

Simplify each expression.

1. $\dfrac{2 - 3}{-4 + 1}$

2. $\dfrac{-2 - 1}{2 + (-3)}$

Evaluate for $x = -1, 0,$ and 2.

3. $f(x) = 3x - 7$

4. $f(x) = 205 + 32x$

DAILY HOMEWORK QUIZ

For use after Lesson 4.8, pages 256–262

1. Does the graph represent a function? Explain your answer.

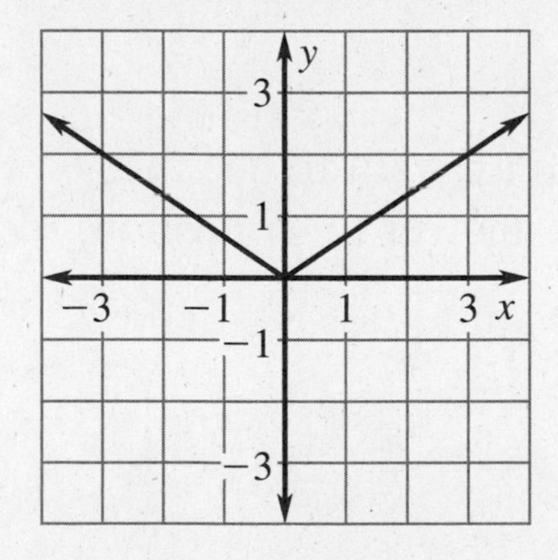

2. Evaluate $f(x) = -3x + 2$ when $x = -1, 0,$ and 3.

3. Find the slope of the graph of the linear function f if $f(3) = -2$ and $f(0) = 1$.

Lesson 5.1

NAME ______________________ DATE __________

Available as a transparency

Graphing Calculator Lesson Opener

For use with pages 273–278

1. a. Graph the equation $y = 2x + 1$.

   b. What is the y-intercept of the line? You may want to use the trace feature or table feature to help you answer this question.

   c. What is the slope of the line? You may want to use the table feature to help you answer this question.

   d. Compare your answers for parts (b) and (c) to the equation in part (a). What do you notice?

2. a. Graph the equation $y = x - 3$.

   b. What is the y-intercept of the line? You may want to use the trace feature or table feature to help you answer this question.

   c. What is the slope of the line? You may want to use the table feature to help you answer this question.

   d. Compare your answers for parts (b) and (c) to the equation in part (a). What do you notice?

3. a. Graph the equation $y = -\frac{1}{2}x + 2$

   b. What is the y-intercept of the line? You may want to use the trace feature or table feature to help you answer this question.

   c. What is the slope of the line? You may want to use the table feature to help you answer this question.

   d. Compare your answers for parts (b) and (c) to the equation in part (a). What do you notice?

4. Consider the equation $y = 5x - 3$. Use your answers to Exercises 1–3 to predict the slope and y-intercept of this equation. Then graph the equation to check your prediction.

NAME ______________________ DATE ________

Practice A

For use with pages 273–278

Find the slope and *y*-intercept of the line.

1. $y = 2x + 5$
2. $y = -4x + 1$
3. $y = x - 5$
4. $y = \frac{1}{2}x$
5. $y = 3 + 2x$
6. $2y = 4x - 3$

Write an equation of the line.

7. The slope is 1; the *y*-intercept is 0.
8. The slope is -2; the *y*-intercept is 4.
9. The slope is -3; the *y*-intercept is -5.
10. The slope is 6; the *y*-intercept is -1.
11. The slope is 0; the *y*-intercept is 9.
12. The slope is -6; the *y*-intercept is -2.
13. The slope is 2; the *y*-intercept is -8.
14. The slope is -4; the *y*-intercept is 11.
15. The slope is 5; the *y*-intercept is 5.
16. The slope is -5; the *y*-intercept is -4.
17. The slope is $-\frac{3}{5}$; the *y*-intercept is 3.
18. The slope is $\frac{8}{9}$; the *y*-intercept is $-\frac{1}{2}$.

Write an equation of the line shown in the graph.

19.
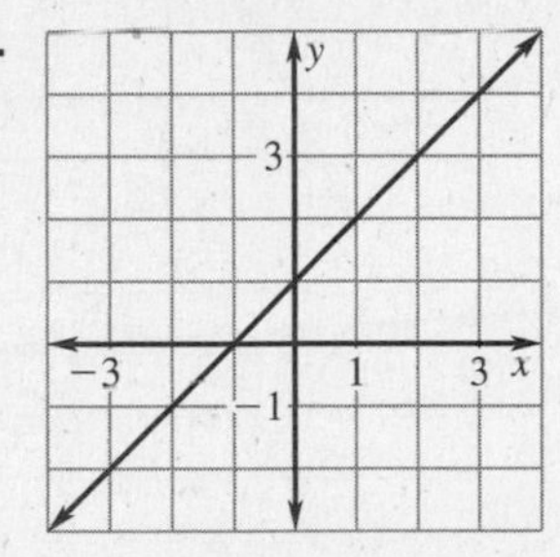

20.
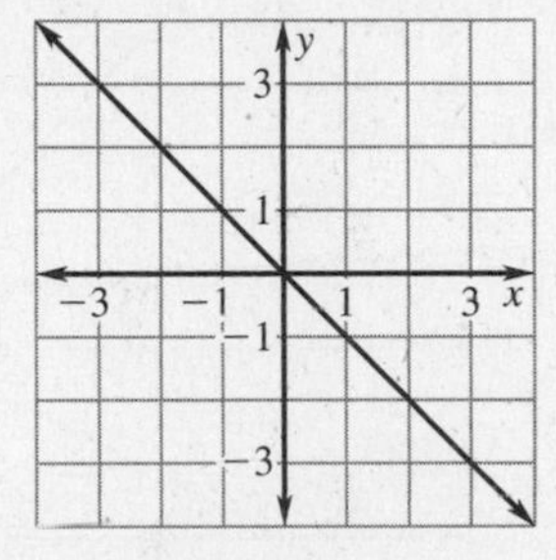

21.
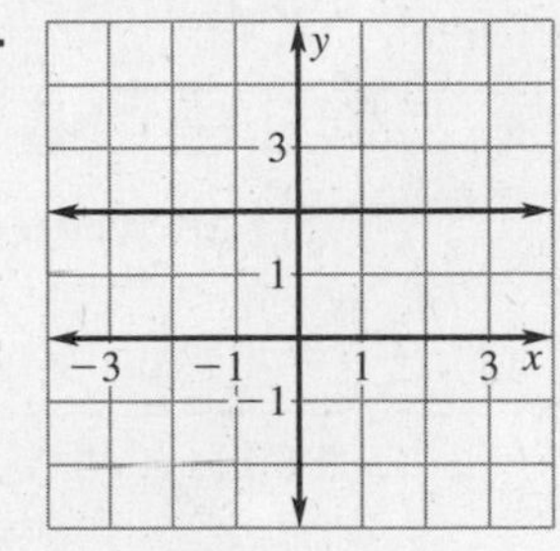

22.
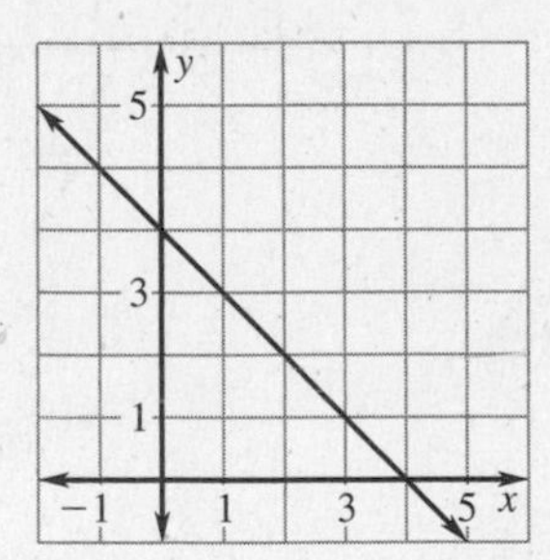

23.
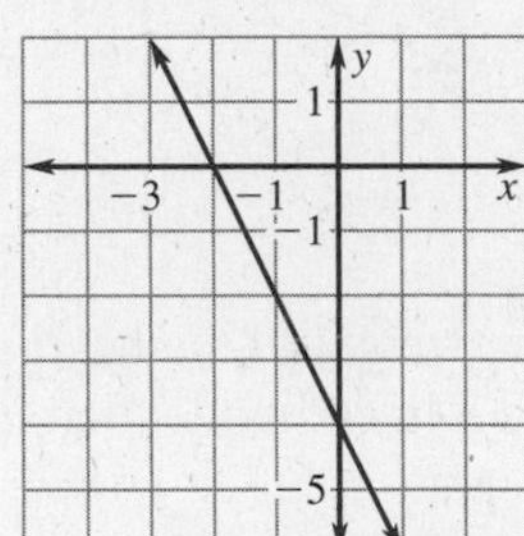

24.
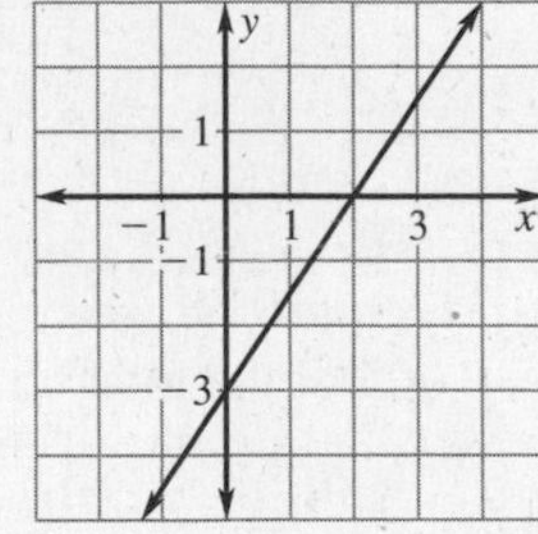

25. ***Child's Height*** Doubling a child's height on his or her second birthday gives a close estimate of his or her final adult height. Write a linear model that gives the approximate adult height of a two-year-old in terms of his or her current height.

26. Use the equation you found in Exercise 25 to complete the table below.

2-year-old height, x (in inches)	31	34	36	37.5
Adult height, y (in inches)				

LESSON 5.1

NAME ______________________________ DATE __________

Practice B

For use with pages 273–278

Write an equation of the line.

1. The slope is 2; the y-intercept is 3.
2. The slope is 5; the y-intercept is 0.
3. The slope is 4; the y-intercept is -3.
4. The slope is -5; the y-intercept is 1.
5. The slope is -3; the y-intercept is -2.
6. The slope is 0; the y-intercept is -5.
7. The slope is $\frac{1}{2}$; the y-intercept is -8.
8. The slope is $-\frac{3}{4}$; the y-intercept is 9.
9. The slope is $-\frac{1}{5}$; the y-intercept is 3.
10. The slope is $\frac{4}{5}$; the y-intercept is -7.
11. The slope is $\frac{1}{3}$; the y-intercept is $\frac{2}{3}$.
12. The slope is $-\frac{4}{3}$; the y-intercept is $\frac{7}{8}$.

Write an equation of the line shown in the graph.

13. 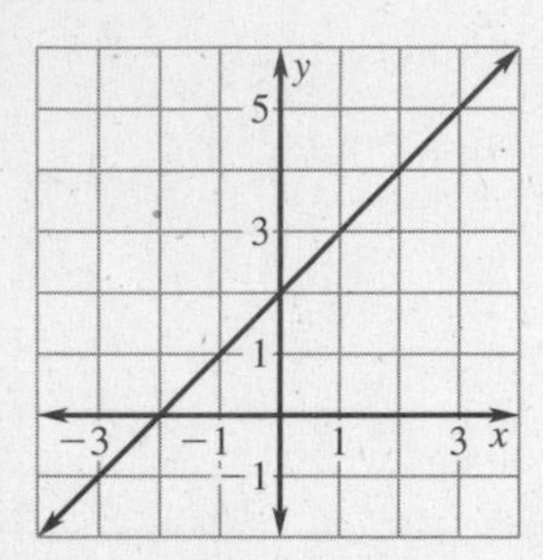

14. 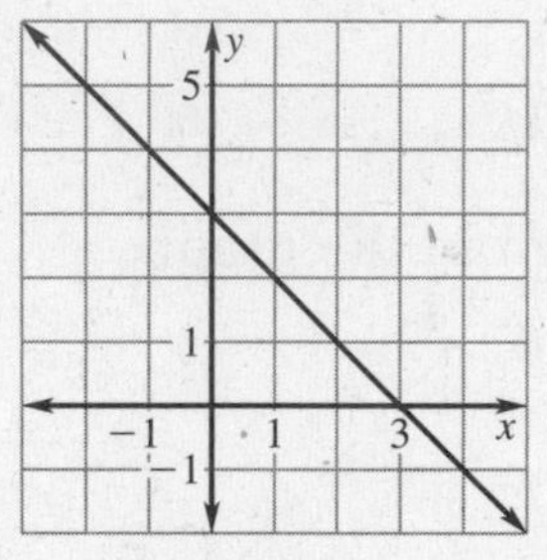

15. 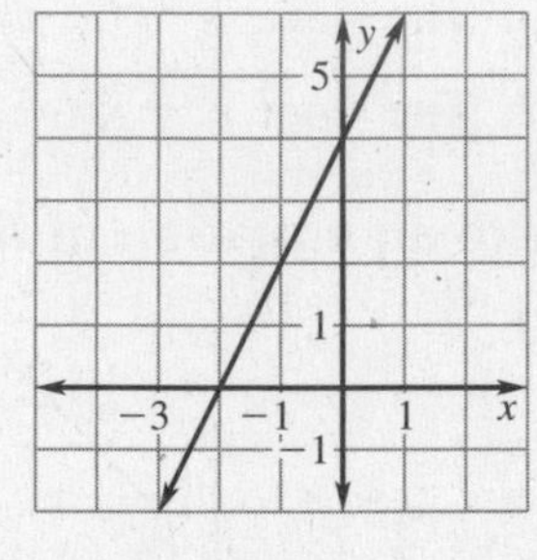

16. 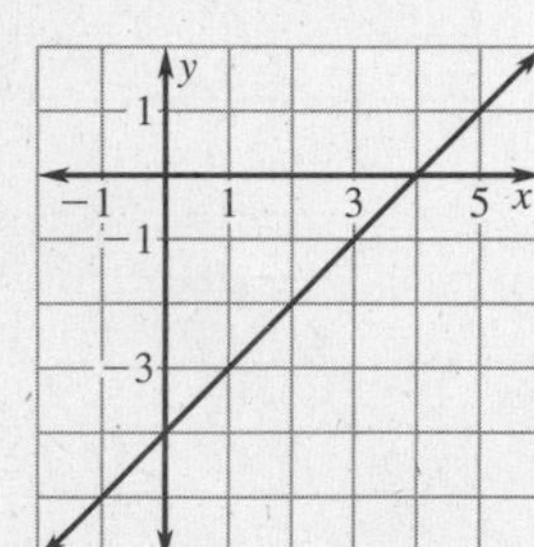

17. 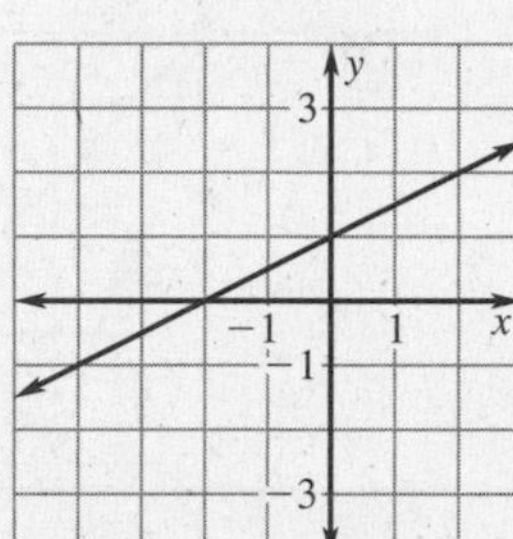

18. 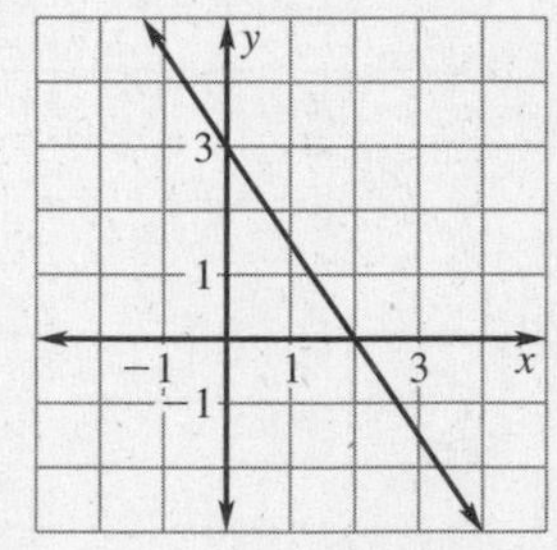

19. *Mammals' Hearts* In mammals, the weight of the heart is approximately 0.005 of the total body weight. Write a linear model that gives the heart weight in terms of the total body weight.

20. Use the equation you found in Exercise 19 to complete the table at the right.

	Human	*Cow*	*Elephant*	*Whale*
Total weight, x (in pounds)	150	1500	12,000	200,000
Heart weight, y (in pounds)				

In Exercises 21 and 22, a car rental company charges a flat fee of \$29 and an additional \$0.15 per mile to rent a compact car.

21. Write an equation to model the total charge, y (in dollars) in terms of x, the number of miles driven.

22. Use the equation you found in Exercise 21 to complete the table at the right.

Miles (x)	25	50	100	200
Cost (y)				

LESSON 5.1

NAME ______________________________ DATE __________

Practice Level C

For use with pages 273–278

Write an equation of the line.

1. The slope is -8; the y-intercept is 5.
2. The slope is 13; the y-intercept is 0.
3. The slope is 1; the y-intercept is -4.
4. The slope is 0; the y-intercept is 7.
5. The slope is $-\frac{1}{4}$; the y-intercept is 3.
6. The slope is $\frac{7}{8}$; the y-intercept is 8.
7. The slope is $\frac{5}{9}$; the y-intercept is -2.
8. The slope is $-\frac{3}{10}$; the y-intercept is 10.
9. The slope is $-\frac{7}{11}$; the y-intercept is $\frac{1}{6}$.
10. The slope is $\frac{9}{16}$; the y-intercept is $-\frac{4}{3}$.

Write an equation of the line shown in the graph.

11. 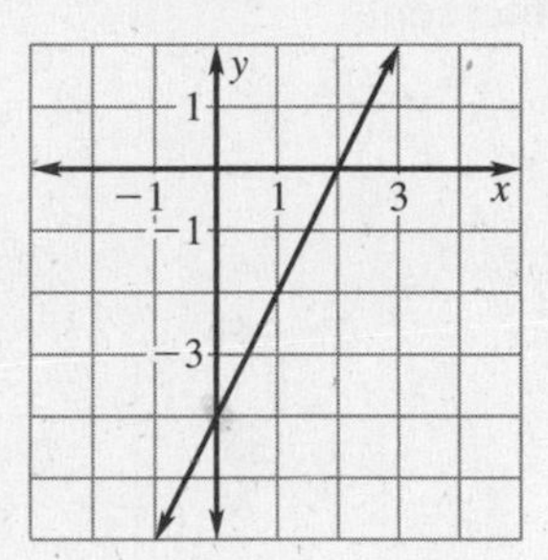

12. 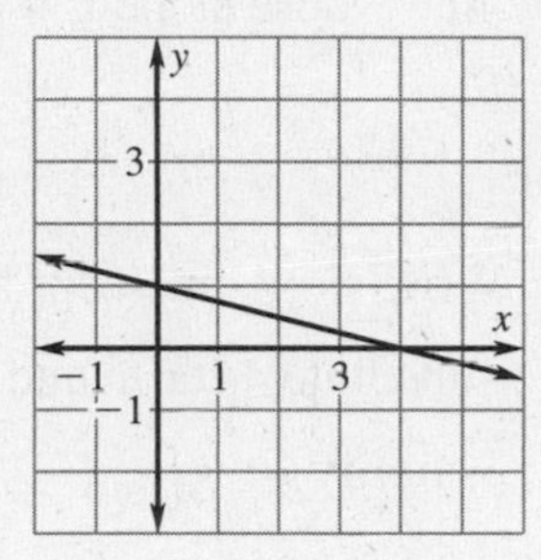

13. 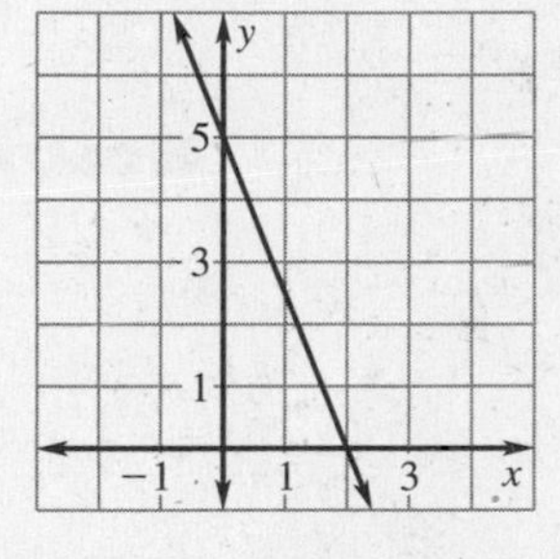

14. 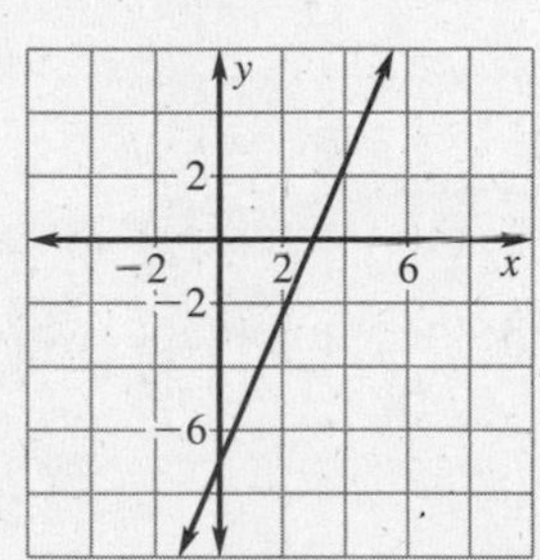

15. 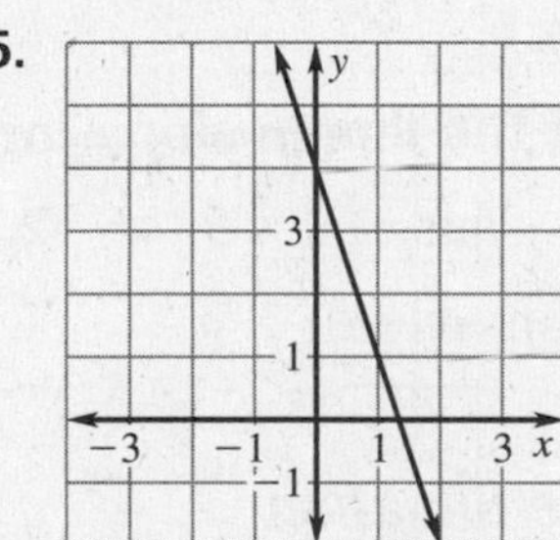

16. 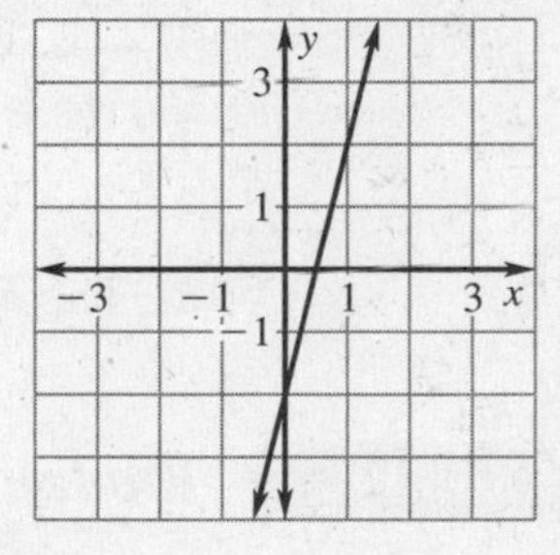

17. ***Photographer's Rate*** A photographer charges \$50 for a sitting and a basic package of photos. Additional 5×7 pictures cost \$12 each. Write a linear model that gives the total cost in dollars in terms of how many extra 5×7 pictures you purchase.

18. Use the equation you found in Exercise 17 to complete the table at the right.

Extra 5×7 ***Photos, x***	0	1	2	3	4
Cost, y					

In Exercises 19–21, a car rental company charges a flat fee of \$31 and an additional \$0.13 per mile to rent a compact car.

19. Write an equation to model the total charge, y (in dollars) in terms of x, the number of miles driven.

20. Use the equation you found in Exercise 19 to complete the table at the right.

Miles (x)	25	50	100	200
Cost (y)				

21. ***Extension*** How would the graph change if each additional mile were \$0.20?

NAME ______________________ DATE __________

Reteaching with Practice

For use with pages 273–278

GOAL **Use the slope-intercept form to write an equation of a line and model a real-life situation with a linear function**

VOCABULARY

In the **slope-intercept form** of the equation of a line, $y = mx + b$, m is the slope and b is the y-intercept.

EXAMPLE 1 *Writing an Equation of a Line*

Write an equation of the line whose slope is 4 and whose y-intercept is -3.

SOLUTION

$y = mx + b$ Write slope-intercept form.

$y = 4x + (-3)$ Substitute 4 for m and -3 for b.

$y = 4x - 3$ Simplify.

Exercises for Example 1

Write an equation of the line in slope-intercept form.

1. The slope is -2; the y-intercept is 5.

2. The slope is 1; the y-intercept is -4.

3. The slope is 0; the y-intercept is 2.

4. The slope is 3; the y-intercept is 6.

EXAMPLE 2 *Modeling a Real-Life Situation*

A car rental company charges a flat fee of \$40 and an additional \$.20 per mile to rent an automobile.

a. Write an equation to model the total charge C (in dollars) in terms of n, the number of miles driven.

b. Complete the table using the equation from part a.

Miles (n)	50	100	200	300
Total charge (C)	?	?	?	?

NAME ______________________________ DATE ____________

Reteaching with Practice

For use with pages 273–278

SOLUTION

a. **Verbal Model** [Total charge] = [Flat fee] + [Rate per mile] · [Number of miles]

Labels Total charge = C (dollars)

Flat fee = 40 (dollars)

Rate per mile = 0.20 (dollars per mile)

Number of miles = n (miles)

Algebraic Model $C = 40 + 0.20 \cdot n$ Linear model

b.

Miles (n)	50	100	200	300
Total charge (C)	50	60	80	100

Exercises for Example 2

5. Rework Example 2 if the company charges a flat fee of $50 and an additional $.30 per mile to rent an automobile.

6. In 1996, the enrollment in your school was approximately 1400 students. During the next three years, the enrollment increased by approximately 30 students per year.

a. Write an equation to model the school's enrollment E in terms of t, the number of years since 1996.

b. Use the equation to estimate the school's enrollment in the year 2002.

LESSON 5.1

NAME ______________________________ DATE ____________

Quick Catch-Up for Absent Students

For use with pages 273–278

The items checked below were covered in class on (date missed) ____________

Lesson 5.1: Writing Linear Equations in Slope-Intercept Form

____ **Goal 1:** Use the slope-intercept form to write an equation of a line. (p. 273)

Material Covered:

____ Example 1: Writing an Equation of a Line

____ Example 2: Writing an Equation of a Line from a Graph

Vocabulary:

slope-intercept form, p. 273

____ **Goal 2:** Model a real-life situation with a linear function. (pp. 274–275)

Material Covered:

____ Student Help: Look Back

____ Example 3: A Linear Model for Population

____ Example 4: A Linear Model for Telephone Charges

____ Other (specify) ____

____

Homework and Additional Learning Support

____ Textbook (specify) pp. 276–278

____

____ *Reteaching with Practice* worksheet (specify exercises) ____________

____ *Personal Student Tutor* for Lesson 5.1

NAME ______________________________ DATE __________

Interdisciplinary Application

For use with pages 273–278

Break-Even Analysis

ECONOMICS One of the most important aspects of business management is determining the price a company will charge for goods and services. Companies will often choose to price products to maximize profits. Another consideration is the cost associated with developing the good or service. One tool used by managers to analyze costs is break-even analysis.

A break-even analysis will determine the quantity of a product that must be sold before the seller begins to make a profit. The analysis takes into consideration variable costs and fixed costs. Variable costs change with the quantity of product produced while fixed costs remain constant. Examples of fixed costs are rent, insurance, administrative salaries, and equipment. Some variable costs are production worker's wages, material expense, and utilities expense.

By graphing both a revenue line and a cost line, a company can then determine a break-even point. This occurs when the revenue and cost lines intersect. This intersection point will be the quantity needed to at least cover costs of production. Any greater quantity will then start generating a profit.

In Exercises 1–4, use the following information.

A store can purchase T-shirts for $7 each. It has fixed costs of $2,500. Each T-shirt is sold for $18.

1. Write a linear equation for both cost and revenue.
2. Graph both the cost line and the revenue line.
3. Determine the break-even point.
4. If the store wants to make a profit of $2,000, how many T-shirts must it sell?

In Exercises 5–7, use the following information.

A hotdog stand can purchase hotdogs for $0.35 each and buns for $0.20 each. It has fixed costs of $50. Each hotdog is sold for $1.

5. Write a linear equation for both cost and revenue.
6. Graph both the cost and revenue lines.
7. Determine the break-even point.

NAME ______________________ DATE ______

Challenge: Skills and Applications

For use with pages 273–278

In Exercises 1–4, use the following information.

According to the census, the population of the United States was about 151 million in 1950 and 249 million in 1990.

1. Find the slope of the line through the two points defined by the population data. What does this slope tell you?
2. Write a linear equation to model the population (in millions) of the United States t years after 1950.
3. Use the equation from Exercise 2 to find the population of the United States in 1970. According to the 1970 census, the population was about 203 million. How close was the amount found with the model to the actual amount? Was this a good approximation? Explain.
4. Use the equation from Exercise 2 to predict the population of the United States in 2010.

In Exercises 5–7, use the following information.

In 1990, people in the United States spent about $285.7 billion on recreation. In 1995, they spent $402.5 billion.

5. Write a linear equation to model the amount (in billions of dollars) spent on recreation t years after 1990.
6. Use the equation from Exercise 5 to find the amount that people in the United States spent on recreation in 1996. The actual amount was $431.1 billion. How close was the amount found with the model to the actual amount? Do you think this is a good approximation? Explain.
7. Use the equation from Exercise 5 to predict the amount people in the United States will spend on recreation in 2000. Do you think the prediction is very accurate? Explain.

LESSON 5.2

TEACHER'S NAME ______________________ CLASS ________ ROOM ________ DATE ________

Lesson Plan

1-day lesson (See *Pacing the Chapter,* TE pages 270C–270D) **For use with pages 279–284**

GOALS
1. **Use slope and any point on a line to write an equation of the line.**
2. **Use a linear model to make predictions about a real-life situation.**

State/Local Objectives ____

____

✓ Check the items you wish to use for this lesson.

STARTING OPTIONS

____ Homework Check: TE page 276; Answer Transparencies
____ Warm-Up or Daily Homework Quiz: TE pages 279 and 278, CRB page 23, or Transparencies

TEACHING OPTIONS

____ Motivating the Lesson: TE page 280
____ Lesson Opener (Activity): CRB page 24 or Transparencies
____ Graphing Calculator Activity with Keystrokes: CRB page 25
____ Examples 1–3: SE pages 279–281
____ Extra Examples: TE pages 280–281 or Transparencies; Internet
____ Closure Question: TE page 281
____ Guided Practice Exercises: SE page 282

APPLY/HOMEWORK

Homework Assignment

____ Basic 12–40 even, 44, 45, 48–50, 55, 65, 75
____ Average 12–40 even, 42–50, 55, 65, 75
____ Advanced 12–40 even, 42–51, 55, 65, 75

Reteaching the Lesson

____ Practice Masters: CRB pages 26–28 (Level A, Level B, Level C)
____ Reteaching with Practice: CRB pages 29–30 or Practice Workbook with Examples
____ Personal Student Tutor

Extending the Lesson

____ Applications (Real-Life): CRB page 32
____ Challenge: SE page 284; CRB page 33 or Internet

ASSESSMENT OPTIONS

____ Checkpoint Exercises: TE pages 280–281 or Transparencies
____ Daily Homework Quiz (5.2): TE page 284, CRB page 36, or Transparencies
____ Standardized Test Practice: SE page 284; TE page 284; STP Workbook; Transparencies

Notes ____

____

____

LESSON 5.2

TEACHER'S NAME ______________________ CLASS ______ ROOM ______ DATE ______

Lesson Plan for Block Scheduling

Half-day lesson (See *Pacing the Chapter,* TE pages 270C–270D) For use with pages 279–284

GOALS
1. **Use slope and any point on a line to write an equation of the line.**
2. **Use a linear model to make predictions about a real-life situation.**

State/Local Objectives ____

CHAPTER PACING GUIDE	
Day	**Lesson**
1	5.1 (all); **5.2 (all)**
2	5.3 (all)
3	5.4 (all); 5.5 (begin)
4	5.5 (end); 5.6 (begin)
5	5.6 (end); 5.7 (begin)
6	5.7 (end); Review Ch. 5
7	Assess Ch. 5; 6.1 (all)

✓ Check the items you wish to use for this lesson.

STARTING OPTIONS

____ Homework Check: TE page 276; Answer Transparencies
____ Warm-Up or Daily Homework Quiz: TE pages 279 and 278, CRB page 23, or Transparencies

TEACHING OPTIONS

____ Motivating the Lesson: TE page 280
____ Lesson Opener (Activity): CRB page 24 or Transparencies
____ Graphing Calculator Activity with Keystrokes: CRB page 25
____ Examples 1–3: SE pages 279–281
____ Extra Examples: TE pages 280–281 or Transparencies; Internet
____ Closure Question: TE page 281
____ Guided Practice Exercises: SE page 282

APPLY/HOMEWORK

Homework Assignment (See also the assignment for Lesson 5.1.)

____ Block Schedule: 12–40 even, 42–50, 55, 65, 75

Reteaching the Lesson

____ Practice Masters: CRB pages 26–28 (Level A, Level B, Level C)
____ Reteaching with Practice: CRB pages 29–30 or Practice Workbook with Examples
____ Personal Student Tutor

Extending the Lesson

____ Applications (Real-Life): CRB page 32
____ Challenge: SE page 284; CRB page 33 or Internet

ASSESSMENT OPTIONS

____ Checkpoint Exercises: TE pages 280–281 or Transparencies
____ Daily Homework Quiz (5.2): TE page 284, CRB page 36, or Transparencies
____ Standardized Test Practice: SE page 284; TE page 284; STP Workbook; Transparencies

Notes ____

NAME ______________________ DATE __________

Available as a transparency

WARM-UP EXERCISES

For use before Lesson 5.2, pages 279–284

Write an equation of the line in slope-intercept form.

1. The slope is 5 and the y-intercept is -2.

2. The slope is $-\frac{2}{3}$ and the y-intercept is 1.

3. The slope is 0 and the y-intercept is $-\frac{1}{2}$.

Solve each equation for *b*.

4. $8 = (-2)(3) + b$

5. $-2 = \frac{1}{4}(-3) + b$

DAILY HOMEWORK QUIZ

For use after Lesson 5.1, pages 273–278

Write an equation of the line in slope-intercept form.

1. slope: $-\frac{3}{2}$; y-intercept: 2

2.

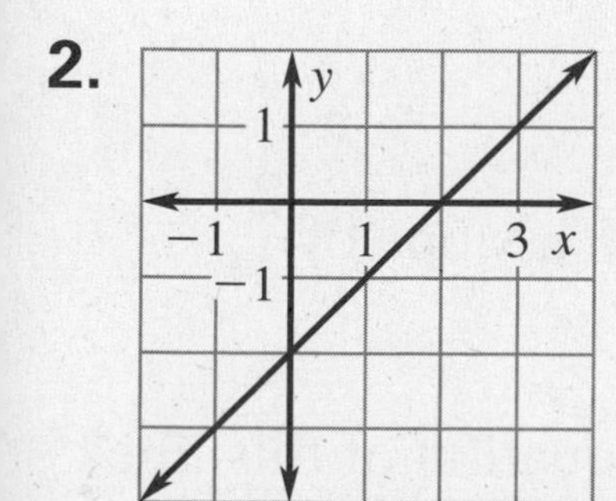

3. You rent a bike for a flat rate of \$15 plus an additional charge of \$2.50 per hour. Write an equation to model the total cost C in terms of the number of hours h you rent the bike.

NAME ______________________________ DATE ____________

Available as a transparency

Activity Lesson Opener

For use with pages 279–284

SET UP: Work with a partner.

YOU WILL NEED: • straightedge • graph paper

1. a. Plot the point $(1, 2)$.

b. Graph the line that passes through $(1, 2)$ with a slope of 2.

c. What is the y-intercept of this line?

d. Do you have enough information to write an equation of this line? Explain. If so, write an equation.

2. a. Plot the point $(-2, 4)$.

b. Graph the line that passes through $(-2, 4)$ with a slope of $-\frac{1}{2}$.

c. What is the y-intercept of this line?

d. Do you have enough information to write an equation of this line? Explain. If so, write an equation.

3. a. Plot the point $(-1, -3)$.

b. Graph the line that passes through $(-1, -3)$ with a slope of -1.

c. What is the y-intercept of this line?

d. Do you have enough information to write an equation of this line? Explain. If so, write an equation.

4. a. Plot the point $(3, -4)$.

b. Graph the line that passes through $(3, -4)$ with a slope of $\frac{2}{3}$.

c. What is the y-intercept of this line?

d. Do you have enough information to write an equation of this line? Explain. If so, write an equation.

5. Make a conjecture about the information you need about a line to write an equation of the line.

NAME ______________________ DATE ____________

Graphing Calculator Activity Keystrokes

For use with page 284

Keystrokes for Example 51

TI-82

Y= .08 X,T,θ + 30 ENTER

40 ENTER

WINDOW ENTER

0 ENTER

300 ENTER

25 ENTER

0 ENTER

200 ENTER

20 ENTER

GRAPH

TI-83

Y= .08 X,T,θ,*n* + 30 ENTER

40 ENTER

WINDOW

0 ENTER

300 ENTER

25 ENTER

0 ENTER

200 ENTER

20 ENTER

GRAPH

SHARP EL-9600c

Step 1:

Y= .08 X/θ/T/*n* + 30 ENTER

40 ENTER

WINDOW

0 ENTER

300 ENTER

25 ENTER

0 ENTER

200 ENTER

20 ENTER

GRAPH

CASIO CFX-9850GA PLUS

From the main menu, choose GRAPH.

.08 X,θ,T + 30 EXE

40 EXE

SHIFT F3

0 EXE

300 EXE

25 EXE

0 EXE

200 EXE

20 EXE

EXIT F6

NAME ____ DATE __________

Practice A

For use with pages 279–284

Write an equation of the line that passes through the point and has the given slope. Write the equation in slope-intercept form.

1. $(0, 2), m = 1$
2. $(-3, 0), m = 4$
3. $(0, 7), m = -5$
4. $(1, 1), m = 3$
5. $(-3, 9), m = 8$
6. $(4, 5), m = -1$
7. $(7, -7), m = -3$
8. $(-4, -15), m = 10$
9. $(5, -10), m = \frac{1}{5}$

Write the slope-intercept form of the equation of the line.

10.
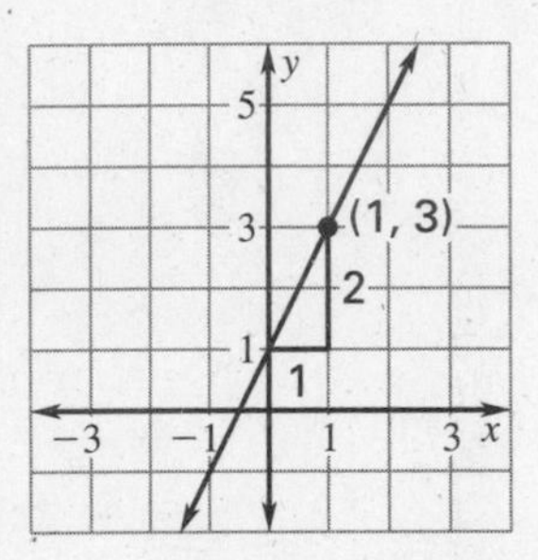

11.
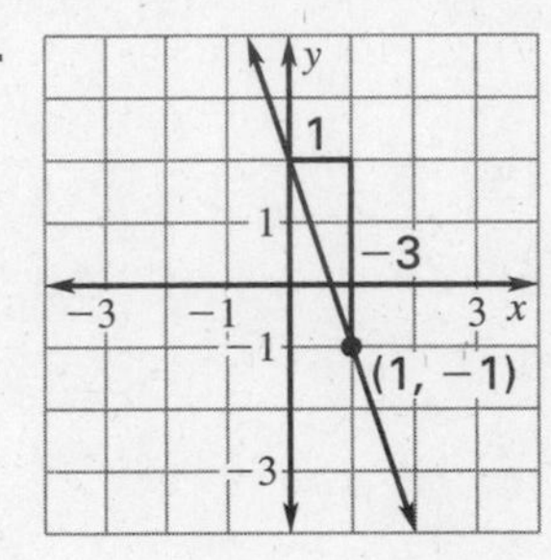

12.
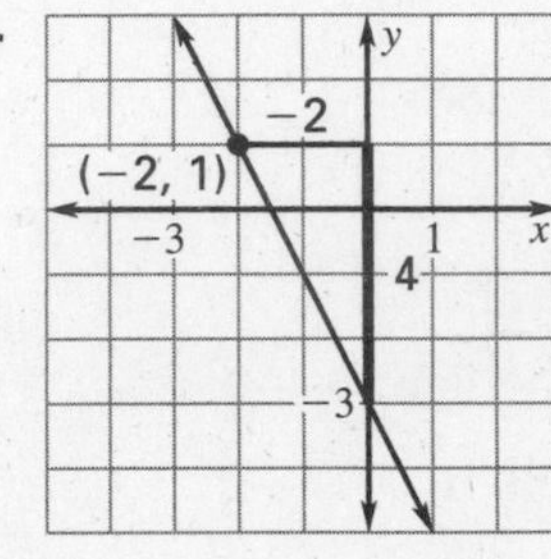

13.
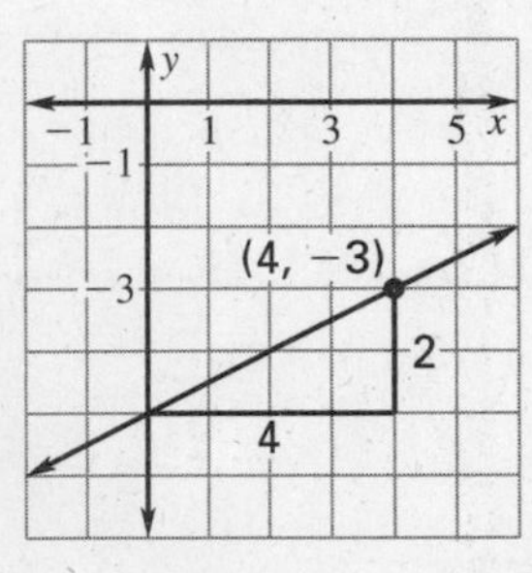

14.
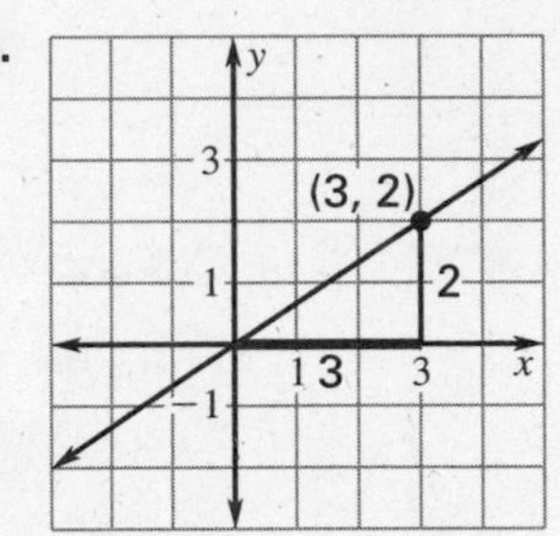

15.
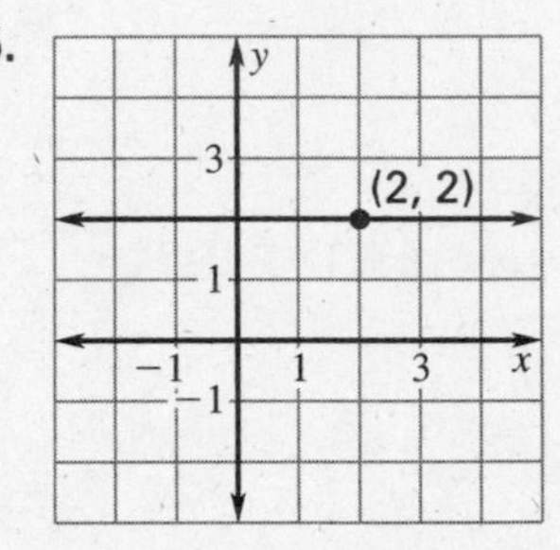

Write an equation of the line that is parallel to the given line and passes through the given point.

16. $y = 3x - 1, (0, 2)$
17. $y = x + 3, (1, 2)$
18. $y = -3x + 5, (-1, 4)$

19. *Apartment Rent* Between 1990 and 2000, the monthly rent for a one-bedroom apartment increased by \$20 per year. In 1997, the rent was \$450 per month. Find an equation that gives the monthly rent in dollars, y, in terms of the year, t. Let $t = 0$ correspond to 1990.

20. *Stamp Collection* Between 1992 and 1999, you added approximately 15 stamps per year to your collection. In 1997 you had 130 stamps. Find an equation that represents the number of stamps in your collection, y, in terms of the year, t. Let $t = 0$ correspond to 1992.

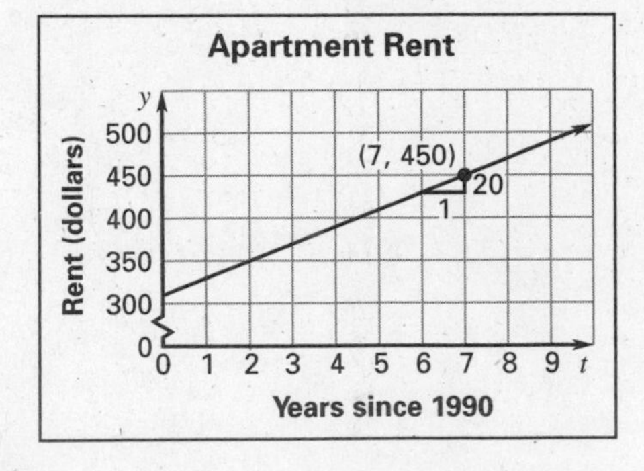

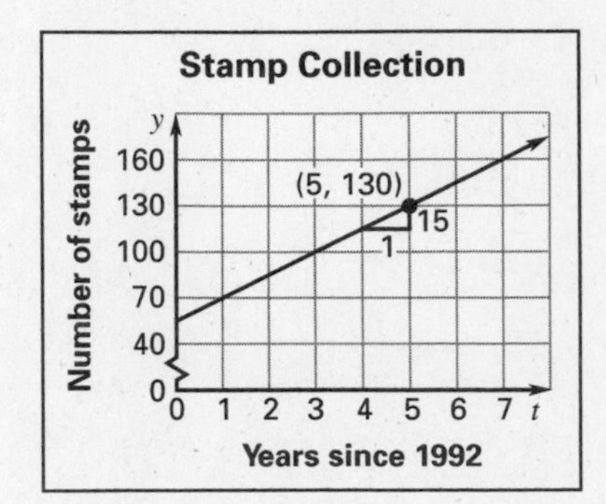

NAME ______________________________ DATE __________

Practice B

For use with pages 279–284

Write an equation of the line that passes through the point and has the given slope. Write the equation in slope-intercept form.

1. $(3, 5), m = -1$
2. $(-2, 6), m = 4$
3. $(7, -2), m = -3$
4. $(2, 8), m = 0$
5. $(-3, 0), m = 2$
6. $(0, 0), m = -7$
7. $(0, -2), m = -\frac{5}{3}$
8. $(-5, -1), m = \frac{3}{4}$
9. $(3, -2), m = -\frac{5}{7}$

Write the slope-intercept form of the equation of the line.

10. 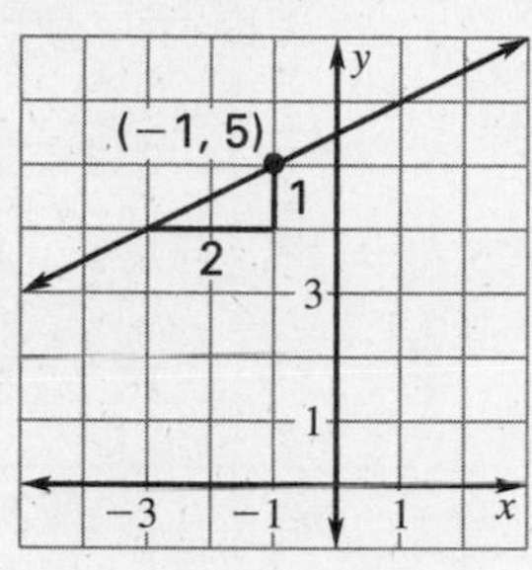

11. 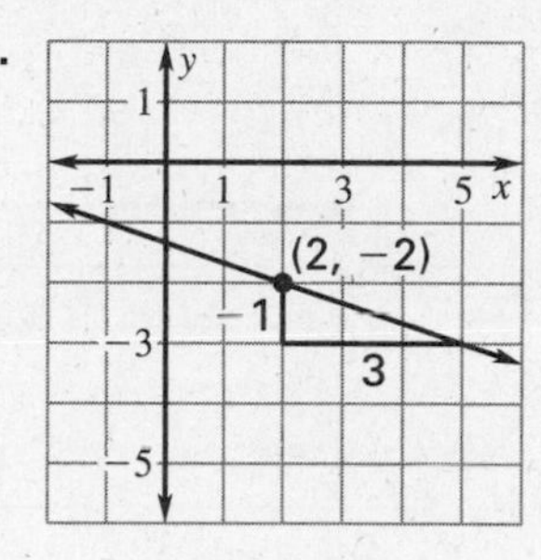

12. 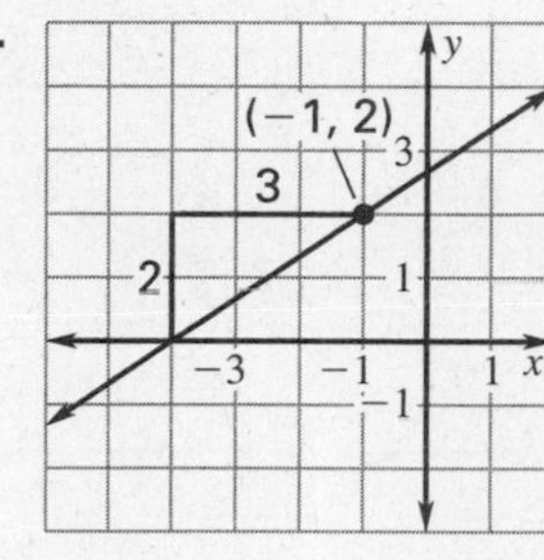

13. 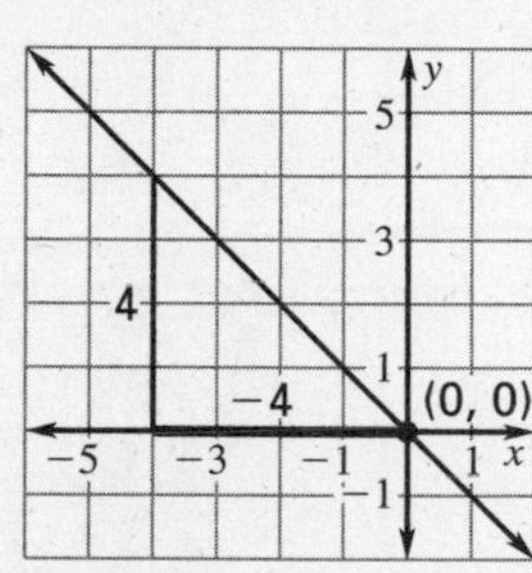

14. 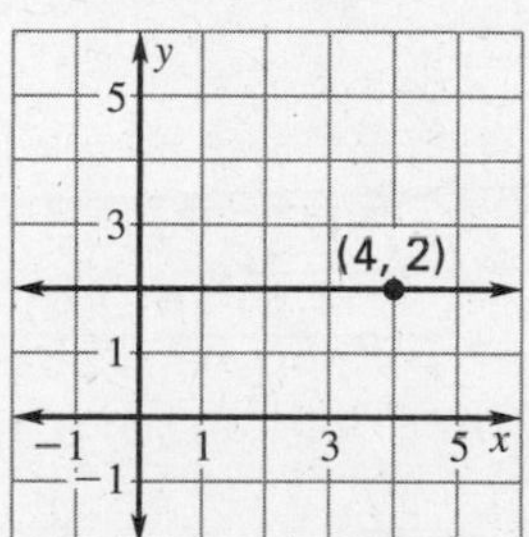

15. 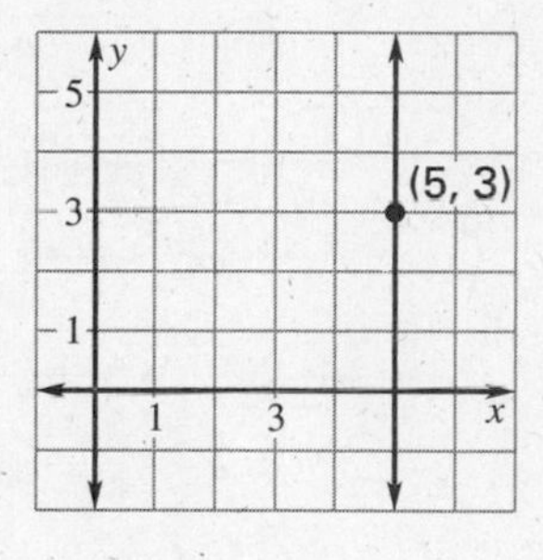

Write an equation of the line that is parallel to the given line and passes through the given point.

16. $y = 5x + 2, (3, 2)$
17. $y = -2x - 1, (2, 6)$
18. $y = \frac{2}{3}x + 5, (1, 1)$

Apartment Rent **In Exercises 19 and 20, use the following information.**

Between 1990 and 2000, the monthly rent for a one-bedroom apartment increased by $27 per year. In 1997, the rent was $375 per month.

19. Find an equation that gives the monthly rent in dollars, y, in terms of the year, t. Let $t = 0$ correspond to 1990.
20. Determine the rent for 1999.

Stamp Collection **In Exercises 21 and 22, use the following information.**

Between 1992 and 1999, you added approximately 21 stamps per year to your collection. In 1997 you had 109 stamps.

21. Find an equation that represents the number of stamps in your collection, y, in terms of the year, t. Let $t = 0$ correspond to 1992.
22. Calculate the number of stamps in 1999.

Salary **In Exercises 23 and 24, use the following information.**

You work as a dental assistant where you are given a $0.75 per hour raise each year. In year three (after two raises), you earn $9.50 per hour.

23. Write an equation that models your hourly wage, w, in terms of the number of years, n, since you started as a dental assistant.
24. What was your starting hourly wage as a dental assistant?

NAME ______________________ DATE __________

Practice C

For use with pages 279–284

Write an equation of the line that passes through the point and has the given slope. Write the equation in slope-intercept form.

1. $(7, 0), m = 4$
2. $(0, -5), m = -\frac{1}{3}$
3. $(-11, -7), m = 1$
4. $(-2, -1), m = 5$
5. $(3, -6), m = \frac{1}{3}$
6. $\left(\frac{3}{4}, -\frac{1}{4}\right), m = 0$
7. $\left(-\frac{5}{7}, 2\right), m = -14$
8. $(-10, 51), m = -\frac{3}{5}$
9. $(-4, -5), m = -\frac{1}{2}$

Write the slope-intercept form of the equation of the line.

10.

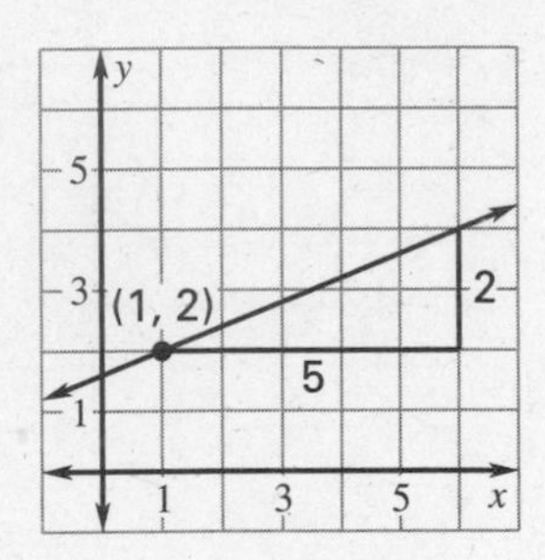

11.

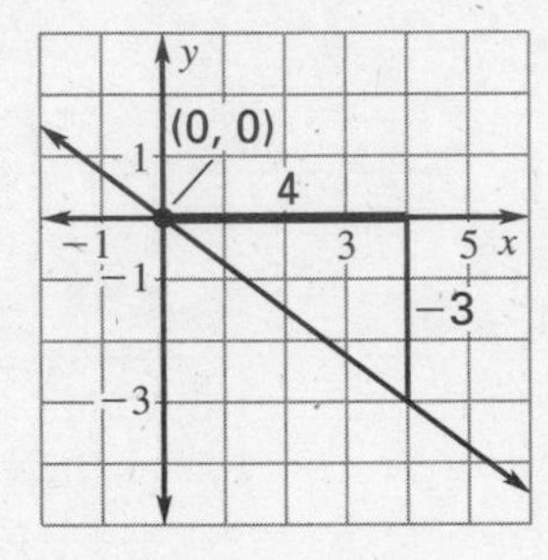

12.

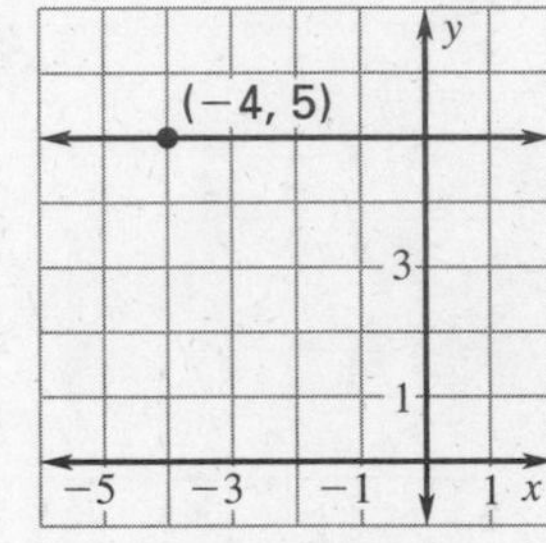

13.

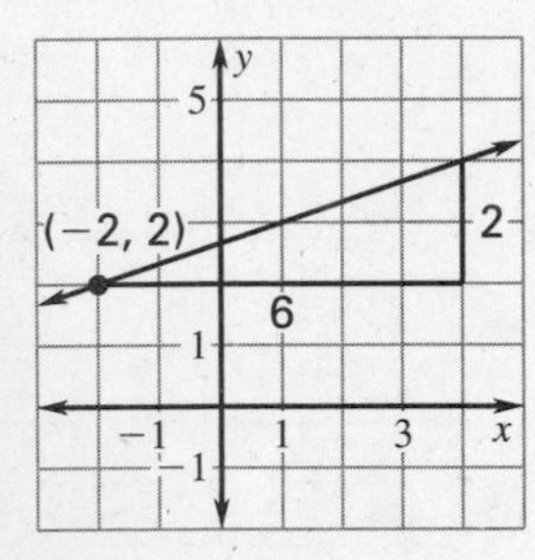

14.

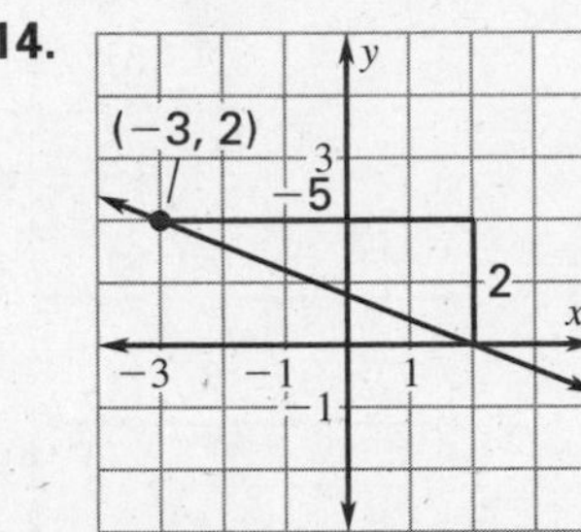

15.

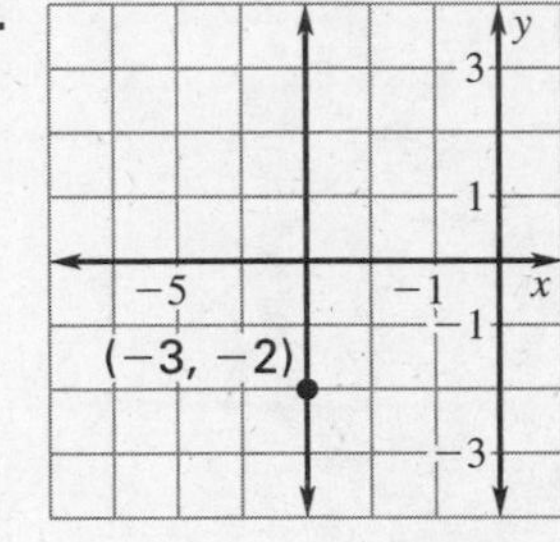

Write an equation of the line that is parallel to the given line and passes through the given point.

16. $2y = 6x - 4, (0, 3)$
17. $3y = x + 5, (6, 0)$
18. $2x - 4y = 8, (-3, 1)$

Apartment Rent **In Exercises 19 and 20, use the following information.**

Between 1990 and 2000, the monthly rent for a one-bedroom apartment increased by \$35 per year. In 1997, the rent was \$420 per month.

19. Find an equation that gives the monthly rent in dollars, y, in terms of the year, t. Let $t = 0$ correspond to 1990.
20. Determine the rent for 1999.

Stamp Collection **In Exercises 21 and 22, use the following information.**

Between 1992 and 1999, you added approximately 35 stamps per year to your collection. In 1997 you had 240 stamps.

21. Find an equation that represents the number of stamps in your collection, y, in terms of the year, t. Let $t = 0$ correspond to 1992.
22. Calculate the number of stamps in 1999.

Salary **In Exercises 23 and 24, use the following information.**
You work as a travel agent and you are given a \$1500 raise at the end of each year. After working three years, you earn \$21,000.

23. Write an equation that models your salary at the end of the year, s, in terms of the number of years, n, since you started as a travel agent.
24. Calculate your salary after working six years as a travel agent.

LESSON 5.2

NAME ______________________ DATE ____________

Reteaching with Practice

For use with pages 279–284

GOAL **Use slope and any point on a line to write an equation of the line and use a linear model to make predictions about a real-life situation**

VOCABULARY

Two nonvertical lines are **parallel** if and only if they have the same slope.

EXAMPLE 1 *Writing an Equation of a Line*

Write an equation of the line that passes through the point $(-2, 5)$ and has a slope of 3.

SOLUTION

Find the y-intercept.

$y = mx + b$	Write slope-intercept form.
$5 = 3(-2) + b$	Substitute 3 for m, -2 for x, and 5 for y.
$5 = -6 + b$	Simplify.
$11 = b$	Solve for b.

The y-intercept is $b = 11$.

Now write an equation of the line, using slope-intercept form.

$y = mx + b$	Write slope-intercept form.
$y = 3x + 11$	Substitute 3 for m and 11 for b.

Exercises for Example 1

Write an equation of the line that passes through the point and has the given slope. Write the equation in slope-intercept form.

1. $(1, -6), m = -2$ **2.** $(-3, -2), m = 4$ **3.** $(4, 5), m = -1$

EXAMPLE 2 *Writing Equations of Parallel Lines*

Write an equation of the line that is parallel to the line $y = 2x + 1$ and passes through the point $(1, 5)$.

SOLUTION

The given line has a slope of $m = 2$. A parallel line through $(1, 5)$ must also have a slope of $m = 2$. Use this information to find the y-intercept.

Lesson 5.2

NAME ______________________________ DATE ____________

Reteaching with Practice

For use with pages 279–284

$y = mx + b$	Write slope-intercept form.
$5 = 2(1) + b$	Substitute 2 for m, 1 for x, and 5 for y.
$5 = 2 + b$	Simplify.
$3 = b$	Solve for b.

The y-intercept is $b = 3$.

Write an equation using the slope-intercept form.

$y = mx + b$	Write slope-intercept form.
$y = 2x + 3$	Substitute 2 for m and 3 for b.

Exercises for Example 2

Write an equation of the line that is parallel to the given line and passes through the given point.

4. $y = 4x - 1,\ (2, 3)$ **5.** $y = x + 6,\ (-3, 0)$ **6.** $y = -2x + 3,\ (1, -1)$

EXAMPLE 3 Writing and Using a Linear Model

The cost of parking in a municipal garage is a base fee plus $1.25 for each hour that you park. Your cost for 5 hours is $10.25. Write a linear equation that models the total cost y of parking in terms of the number of hours x.

SOLUTION

The slope is 1.25 and $(x, y) = (5, 10.25)$ is a point on the line.

$y = mx + b$	Write slope-intercept form.
$10.25 = (1.25)(5) + b$	Substitute 1.25 for m, 5 for x, and 10.25 for y.
$10.25 = 6.25 + b$	Simplify.
$4 = b$	The y-intercept is $b = 4$.

Write an equation of the line using $m = 1.25$ and $b = 4$.

$y = mx + b$	Write slope-intercept form.
$y = 1.25x + 4$	Substitute 1.25 for m and 4 for b.

Exercise for Example 3

7. Use the linear equation from Example 3 to estimate the total cost y of parking for 7 hours.

LESSON 5.2

NAME ______________________________ DATE __________

Quick Catch-Up for Absent Students

For use with pages 279–284

The items checked below were covered in class on (date missed) ____________

Lesson 5.2: Writing Linear Equations Given the Slope and a Point

____ **Goal 1:** Use slope and any point on a line to write an equation of the line. (pp. 279–280)

Material Covered:

____ Example 1: Writing an Equation of a Line

____ Student Help: Look Back

____ Example 2: Writing Equations of Parallel Lines

____ **Goal 2:** Use a linear model to make predictions about a real-life situation. (p. 281)

Material Covered:

____ Example 3: Writing and Using a Linear Model

____ Other (specify) ______________________________

____

Homework and Additional Learning Support

____ Textbook (specify) pp. 282–284

____

____ Internet: Extra Examples at www.mcdougallittell.com

____ *Reteaching with Practice* worksheet (specify exercises) ____________

____ *Personal Student Tutor* for Lesson 5.2

NAME ______________________________ DATE __________

Real-Life Application: When Will I Ever Use This?

For use with pages 279–284

Sports Participation

In most of the world, when a person says "football" they are actually referring to what Americans call soccer. Soccer is the most universally played sport in the world. Both soccer and American football have common origins. The first official rules of soccer were written in 1863.

Colleges organized some of the first football games and the first intercollegiate game was on November 6, 1869 between Princeton and Rutgers. It was a combination of soccer and football. Walter Camp, a Yale coach, is known as the father of American football.

Your school board is distributing next year's sports budget to each individual sport based on projected participation. In 1999, boys' soccer had 142 members and has increased 12 students per year since 1990. Girls' soccer had 120 members and has increased 8 students per year since 1990. Football has 92 members and has decreased by 3 students per year since 1990.

1. Write a linear equation for each sport that models the number of students y in terms of x, the number of years since 1990.
2. Graph each equation in the same coordinate plane, labeling each line.
3. Estimate the number of students in each sport in 2001.
4. In the year 2001 the budget is $250,000. Estimate how much funding each sport would receive in 2001.

NAME ______________________ DATE __________

Challenge: Skills and Applications

For use with pages 279–284

In Exercises 1–4, a line with the given slope *m* contains the given point. Find the *y*-intercept of the line.

1. $m = \frac{2}{3}$; $(-12, 1)$

2. $m = -\frac{5}{2}$; $(6, -4)$

3. $m = \frac{3}{4}$; $(-2, 7)$

4. $m = -\frac{5}{6}$; $(4, -8)$

5. A line with a slope of -3 passes through the point $(k, 4)$. Find the y-intercept of the line in terms of k.

6. A line with slope of $-\frac{1}{r}$ passes through the point (h, k). Find the y-intercept in terms of r, h, and k.

7. Suppose a certain line has equation $y = mx + b$ and passes through the point $(4, q)$. Suppose another line has the same y-intercept and passes through the point $(4, q + 2)$. Write an equation of this second line using the variables m and b from the first equation, but not using the variable q.

In Exercises 8–9, write an equation of the line.

8. the line whose slope is the same as that of the line $2x - 3y = 4$ and that passes through $(1, 7)$

9. the line with slope p passing through (p, q)

In Exercises 10–12, use the following information.

David Margolez has been working for Pioneer Engineering since 1990. Each year he gets a \$2100 raise. In 1998, he earned one and a half times as much as he earned in 1990.

10. Write an equation in the form $y = mx + b$ that models David's salary y in terms of the number of years x since he started working at Pioneer Engineering. (*Hint*: Think about what the slope m and the y-intercept b represent in this problem and what the given information tells you about them.)

11. What was David's salary in 1998?

12. Use your equation to predict how much David will earn in 2005.

LESSON
5.3

TEACHER'S NAME ________________ CLASS ______ ROOM ______ DATE ______

Lesson Plan

2-day lesson (See *Pacing the Chapter,* TE pages 270C–270D) For use with pages 285–291

GOALS
1. **Write an equation of a line given two points on the line.**
2. **Use a linear equation to model a real-life problem.**

State/Local Objectives ____

____

✓ Check the items you wish to use for this lesson.

STARTING OPTIONS
____ Homework Check: TE page 282; Answer Transparencies
____ Warm-Up or Daily Homework Quiz: TE pages 285 and 284, CRB page 36, or Transparencies

TEACHING OPTIONS
____ Motivating the Lesson: TE page 286
____ Lesson Opener (Application): CRB page 37 or Transparencies
____ Graphing Calculator Activity with Keystrokes: CRB pages 38–39
____ Examples: Day 1: 1–2, SE pages 285–286; Day 2: 3, SE page 287
____ Extra Examples: Day 1: TE page 286 or Transp.; Day 2: TE page 287 or Transp.; Internet
____ Closure Question: TE page 287
____ Guided Practice: SE page 288; Day 1: Exs. 1–17; Day 2: none

APPLY/HOMEWORK
Homework Assignment
____ Basic Day 1: 18–44 even; Day 2: 45–52, 58–60, 66–76 even; Quiz 1: 1–13
____ Average Day 1: 18–44 even; Day 2: 45–55, 58–60, 66–76 even; Quiz 1: 1–13
____ Advanced Day 1: 18–44 even; Day 2: 45–47, 51–65, 66–76 even; Quiz 1: 1–13

Reteaching the Lesson
____ Practice Masters: CRB pages 40–42 (Level A, Level B, Level C)
____ Reteaching with Practice: CRB pages 43–44 or Practice Workbook with Examples
____ Personal Student Tutor

Extending the Lesson
____ Applications (Interdisciplinary): CRB page 46
____ Challenge: SE page 290; CRB page 47 or Internet

ASSESSMENT OPTIONS
____ Checkpoint Exercises: Day 1: TE page 286 or Transp.; Day 2: TE page 287 or Transp.
____ Daily Homework Quiz (5.3): TE page 290, CRB page 51, or Transparencies
____ Standardized Test Practice: SE page 290; TE page 290; STP Workbook; Transparencies
____ Quiz (5.1–5.3): SE page 291; CRB page 48

Notes ____

____

____

LESSON 5.3

TEACHER'S NAME ______________________ CLASS ________ ROOM ________ DATE ________

Lesson Plan for Block Scheduling

1-day lesson (See *Pacing the Chapter,* TE pages 270C–270D) **For use with pages 285–291**

GOALS
1. **Write an equation of a line given two points on the line.**
2. **Use a linear equation to model a real-life problem.**

State/Local Objectives ______________________

CHAPTER PACING GUIDE	
Day	**Lesson**
1	5.1 (all); 5.2 (all)
2	**5.3 (all)**
3	5.4 (all); 5.5 (begin)
4	5.5 (end); 5.6 (begin)
5	5.6 (end); 5.7 (begin)
6	5.7 (end); Review Ch. 5
7	Assess Ch. 5; 6.1 (all)

✓ Check the items you wish to use for this lesson.

STARTING OPTIONS

____ Homework Check: TE page 282: Answer Transparencies
____ Warm-Up or Daily Homework Quiz: TE pages 285 and 284, CRB page 36, or Transparencies

TEACHING OPTIONS

____ Motivating the Lesson: TE page 286
____ Lesson Opener (Application): CRB page 37 or Transparencies
____ Graphing Calculator Activity with Keystrokes: CRB pages 38–39
____ Examples 1–3: SE pages 285–287
____ Extra Examples: TE pages 286–287 or Transparencies; Internet
____ Closure Question: TE page 287
____ Guided Practice Exercises: SE page 288

APPLY/HOMEWORK

Homework Assignment

____ Block Schedule: 18–44 even, 45–55, 58–60, 66–76 even; Quiz 1: 1–13

Reteaching the Lesson

____ Practice Masters: CRB pages 40–42 (Level A, Level B, Level C)
____ Reteaching with Practice: CRB pages 43–44 or Practice Workbook with Examples
____ Personal Student Tutor

Extending the Lesson

____ Applications (Interdisciplinary): CRB page 46
____ Challenge: SE page 290; CRB page 47 or Internet

ASSESSMENT OPTIONS

____ Checkpoint Exercises: TE pages 286–287 or Transparencies
____ Daily Homework Quiz (5.3): TE page 290, CRB page 51, or Transparencies
____ Standardized Test Practice: SE page 290; TE page 290; STP Workbook; Transparencies
____ Quiz (5.1–5.3): SE page 291; CRB page 48

Notes ______________________

LESSON 5.3

NAME ______________________ DATE ________

Available as a transparency

Warm-Up Exercises

For use before Lesson 5.3, pages 285–291

Write the equation in slope-intercept form of the line that passes through the point and has the given slope.

1. $(-2, 5), m = \frac{1}{2}$

2. $(6, -3), m = -2$

Write an equation of the line that is parallel to the given line and passes through the given point.

3. $y = -3x + 2,\ (2, 1)$

4. $y = \frac{1}{4}x + 1,\ (-2, 0)$

Daily Homework Quiz

For use after Lesson 5.2, pages 279–284

Write an equation of the line that passes through the point and has the given slope.

1. a. $(6, -2),\ m = -1$

b. $(2, 0),\ m = 4$

2. Write an equation of the line through $(4, -1)$ that is parallel to the line $y = 2x - 3$.

3. Between 1990 and 1998, the population of Riverton decreased by about 150 people per year. In 1997, the population was 75,450.

a. Write a linear equation that models the population P of Riverton in terms of the year t, where t is the number of years since 1990.

b. Estimate Riverton's population in 2003.

NAME ______________________ DATE __________

Available as a transparency

Application Lesson Opener

For use with pages 285–291

You have a job that pays by the hour. On Monday, you worked 5 hours and earned $30. On Tuesday, you worked 3 hours and earned $18.

1. Let x represent the number of hours worked and let y represent the amount earned. Write two ordered pairs.

2. Plot the two points given by the ordered pairs. Draw a line through the points.

3. Find the slope of the line.

4. Find the y-intercept of the line.

You are an Internet provider that charges $10 a month plus an hourly fee to use the Internet. One month you paid $4 for 2 hours of Internet time, for a total monthly bill of $14. Another month, you paid $10 for 5 hours of Internet time, for a total monthly bill of $20.

5. Let x represent the number of hours and let y represent the total monthly bill. Write two ordered pairs.

6. Plot the two points given by the ordered pairs. Draw a line through the points.

7. Find the slope of the line.

8. Find the y-intercept of the line.

9. In the two situations above, do you have enough information to write equations of the lines? Explain your answer.

LESSON 5.3

NAME ______________________ DATE ________

Graphing Calculator Activity

For use with pages 285–291

GOAL **To determine whether two different lines in the same plane are perpendicular**

Geometrically, perpendicular lines intersect to form a right angle. Algebraically, the slopes of perpendicular lines also relate. A graphing calculator can be used to visually check whether two lines are perpendicular and to discover this slope relationship.

Activity

1. Enter the following equations into your graphing calculator.

   $y = \frac{4}{5}x + 2$ $\qquad$ $y = -\frac{4}{5}x + 2$

2. Plot the graph of each equation in the same coordinate plane. Do the lines appear perpendicular?
3. Compare the slopes of the equations.
4. Repeat Steps 1–3 with each pair of equations. Be sure to clear out the equations from Step 1.

   a. $y = \frac{4}{5}x + 2,\ y = \frac{5}{4}x + 2$ $\qquad$ **b.** $y = \frac{4}{5}x + 2,\ y = -\frac{5}{4}x + 2$

5. Write a statement about the slopes of perpendicular lines.

Exercises

1. Determine the unknown slope that would make the lines perpendicular. Then use your graphing calculator to visually check your answer.

a. $y = \frac{1}{3}x$

$y = \underline{\ ?\ }x$

b. $y = -2x + 3$

$y = \underline{\ ?\ }x$

c. $y = \frac{3}{2}x - 5$

$y = \underline{\ ?\ }x$

2. Start with the equation $y = \frac{1}{2}x$. Write three other linear equations that, along with the first equation, will create a rectangle. Use your graphing calculator to visually check your answer.

See page 39 for keystrokes.

NAME ______________________________ DATE __________

Graphing Calculator Activity

For use with pages 285–291

TI-82

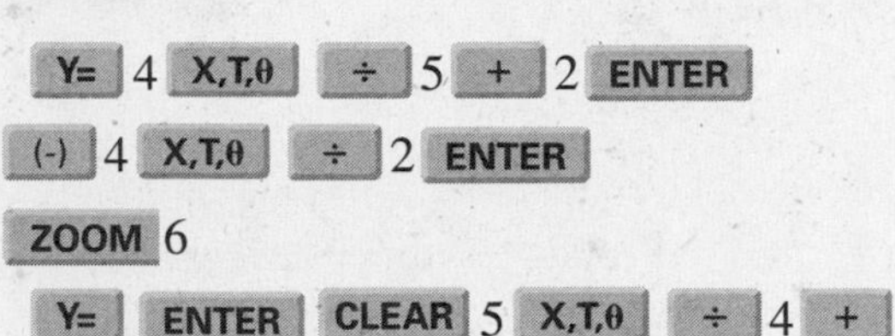

TI-83

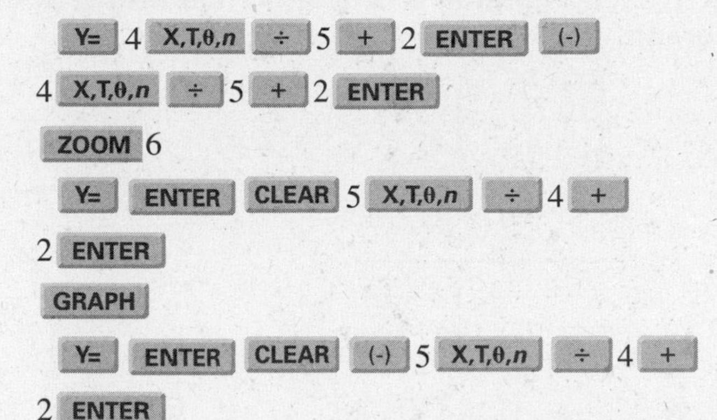

SHARP EL-9600c

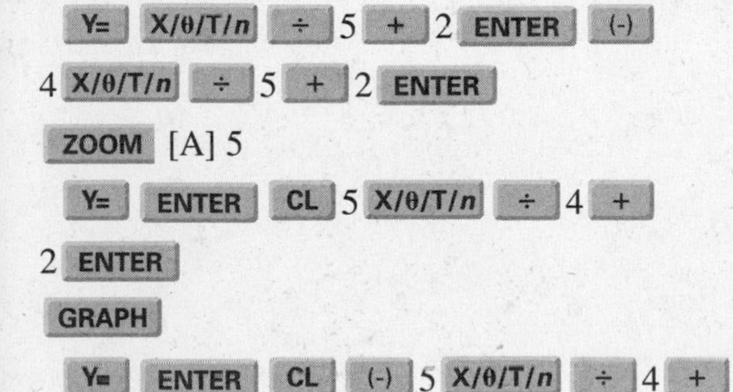

CASIO CFX-9850GA PLUS

From the main menu, choose GRAPH.

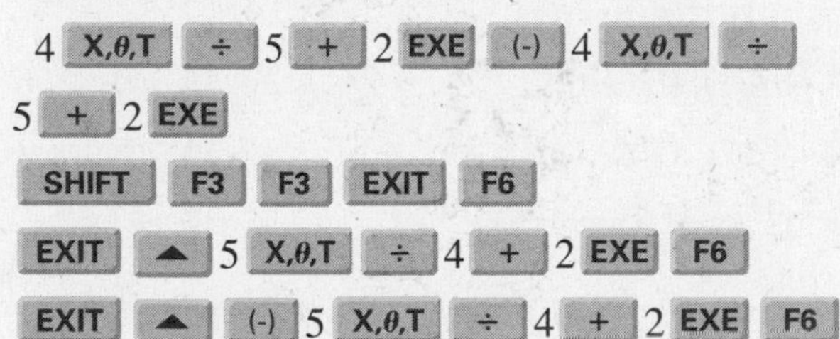

LESSON **5.3**

NAME ______________________ DATE __________

Practice A

For use with pages 285–291

Write an equation in slope-intercept form of the line shown in the graph.

1.

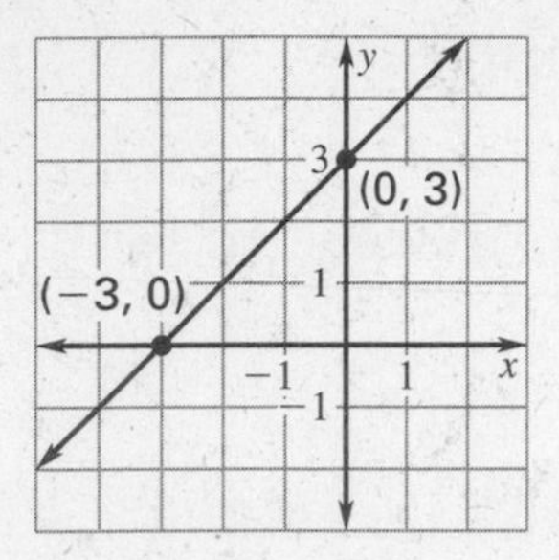

2.

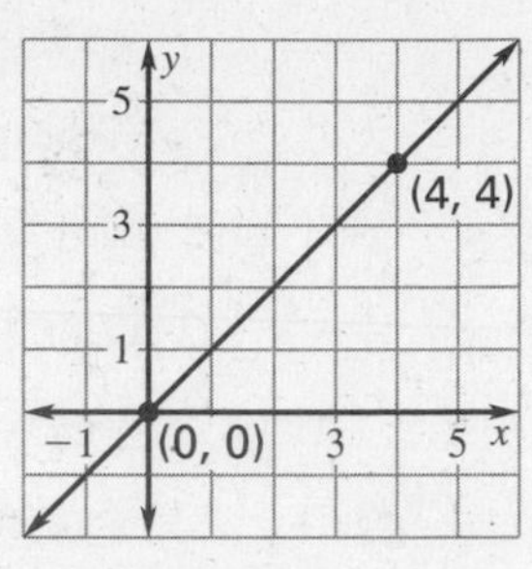

3.

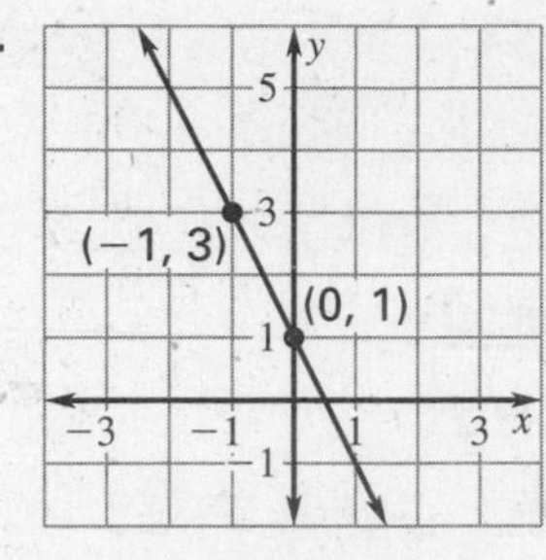

4.

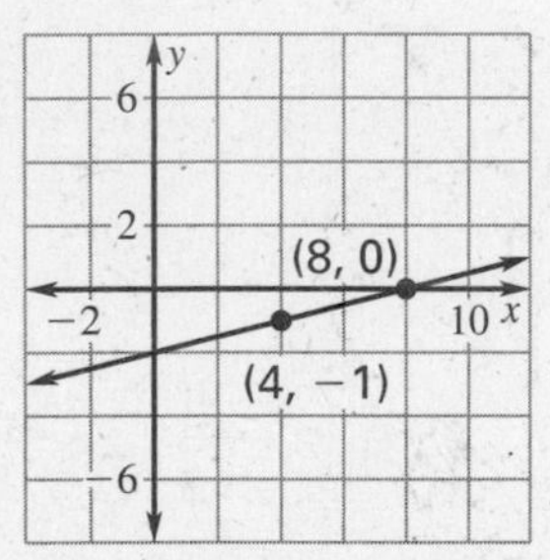

5.

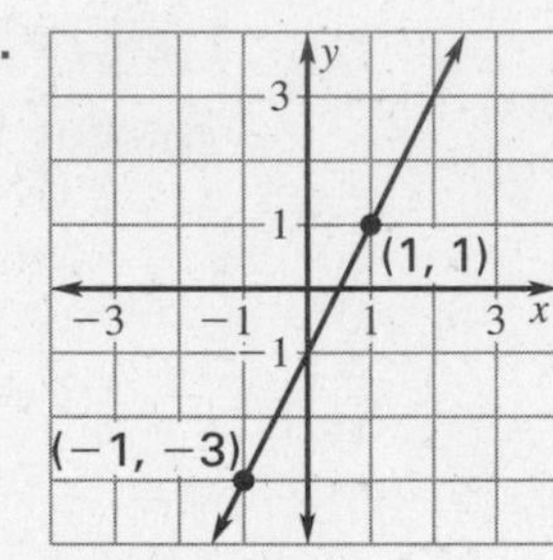

6.

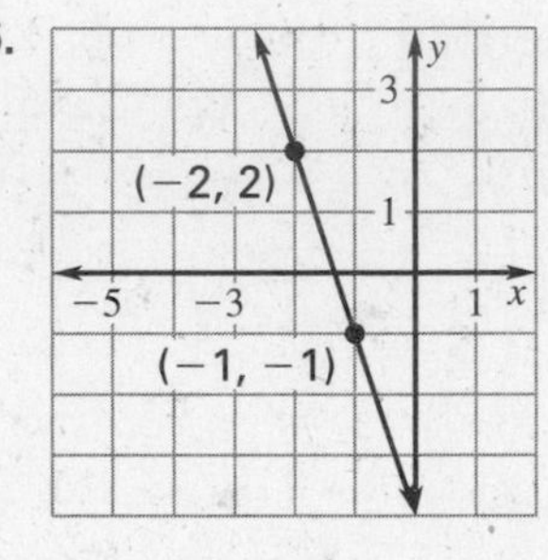

Write an equation in slope-intercept form of the line that passes through the points.

7. $(0, 0), (3, -6)$

8. $(0, 4), (-1, 3)$

9. $(-5, 9), (-2, 0)$

10. $(0, 2), (-2, 0)$

11. $(5, 0), (-10, -5)$

12. $(1, 1), (3, 3)$

13. $(1, -7), (3, -15)$

14. $(-6, -2), (-10, -14)$

15. $(2, 3), (6, 11)$

Give the slope of a line perpendicular to the given line.

16. $y = x - 2$

17. $y = -3x + 9$

18. $y = \frac{1}{2}x + 4$

19. ***Learning a Language*** By the end of your 5th French lesson you have learned 20 vocabulary words. Write an equation that gives the number of vocabulary words you know, y, in terms of the number of lessons you have had, x. Assume that you learn the same number of words at each lesson.

20. ***United Nations*** In 1945, when the United Nations was formed, there were 51 member nations. In 1987, there were 159 member nations. Write an equation that gives the number of nations in the UN, y, in terms of the year, t. Let $t = 0$ correspond to 1945 and assume that membership followed a linear pattern.

21. ***Diving*** Leslie dives off a block at the edge of the pool. She enters the water 8 ft from the side of the pool. Leslie is 1 ft under water when she is 11 ft from the side of the pool. Write an equation that gives Leslie's depth, y, in terms of her distance from the side.

22. ***Nature Hike*** Use the diagram at the right to write the equation of the line from point A to point B. What is the slope of this line?

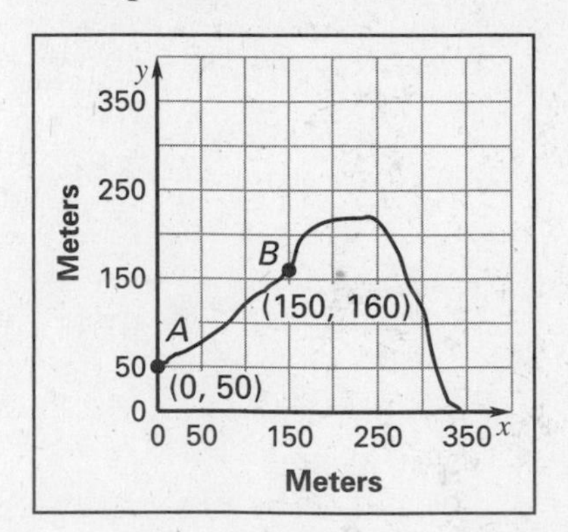

Lesson 5.3

NAME ____ DATE ___________

Practice B

For use with pages 285–291

Write an equation in slope-intercept form of the line shown in the graph.

1.

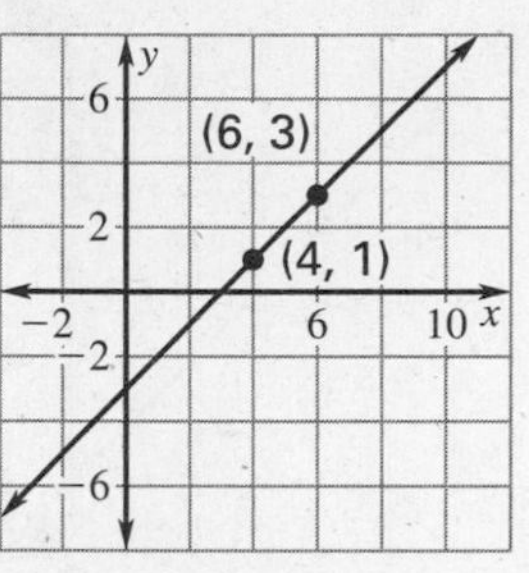

2.

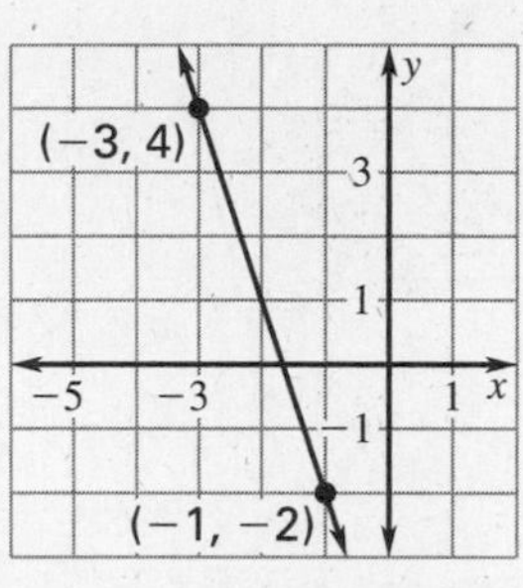

3.

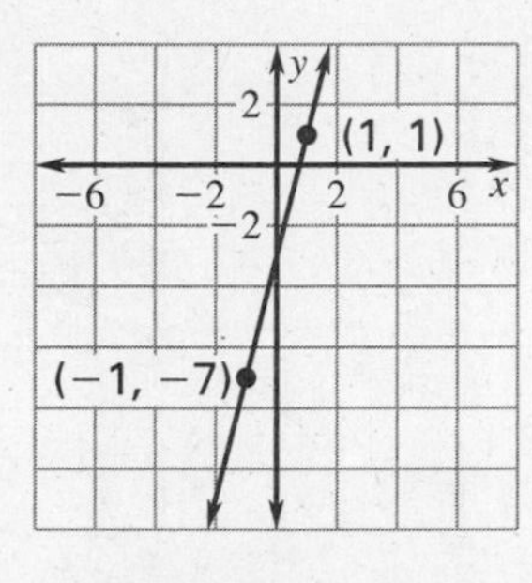

4.

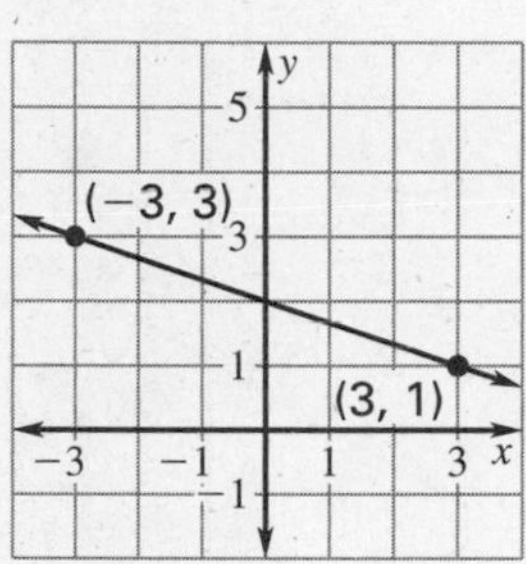

5.

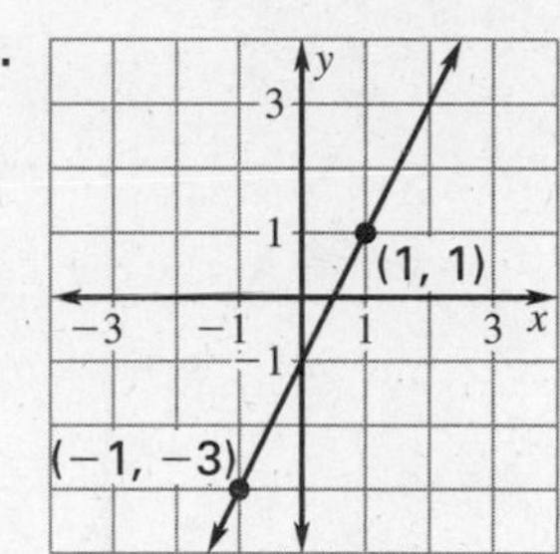

6.

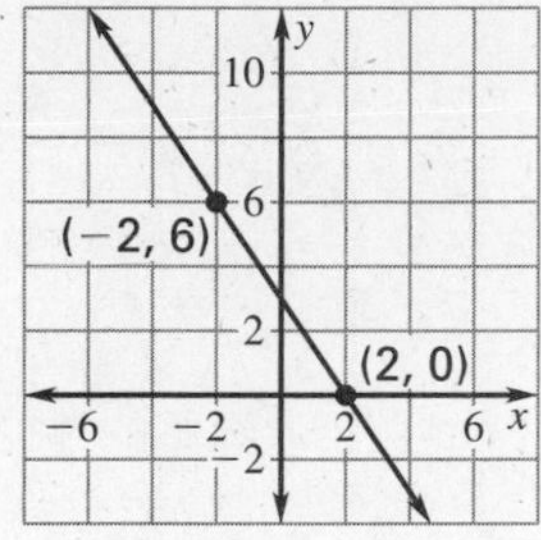

Write an equation in slope-intercept form of the line that passes through the points.

7. $(0, 8), (-1, 3)$
8. $(-7, 9), (-5, -3)$
9. $(3, 2), (7, 5)$
10. $(4, 2), (3, 5)$
11. $(-5, -6), (2, 8)$
12. $(-5, 6), (-6, 1)$
13. $\left(\frac{1}{2}, -1\right), \left(3, \frac{3}{2}\right)$
14. $(6.22, -3.75), (-1.78, 0.25)$
15. $\left(\frac{1}{8}, \frac{7}{8}\right), \left(\frac{3}{4}, -\frac{5}{4}\right)$

Give the slope of a line perpendicular to the given line.

16. $y = 3x + 5$
17. $y = -\frac{2}{3}x - 4$
18. $y = -2x + 6$

Geometry Connection **In Exercises 19–21, use the graph.**

19. Find the perpendicular sides of trapezoid $WXYZ$. How do you know mathematically that these two sides are perpendicular?

20. Write equations of the lines passing through the perpendicular sides.

21. Write equations of the lines passing through the two parallel sides. How do you know mathematically that these two sides are parallel?

22. ***Driving*** You drove to your cousin's house, which is 460 miles away. After two hours, you had gone 100 miles. After 8 hours, you reached your destination. Write an equation that gives the number of miles you had driven, y, in terms of the number of hours you had driven, t.

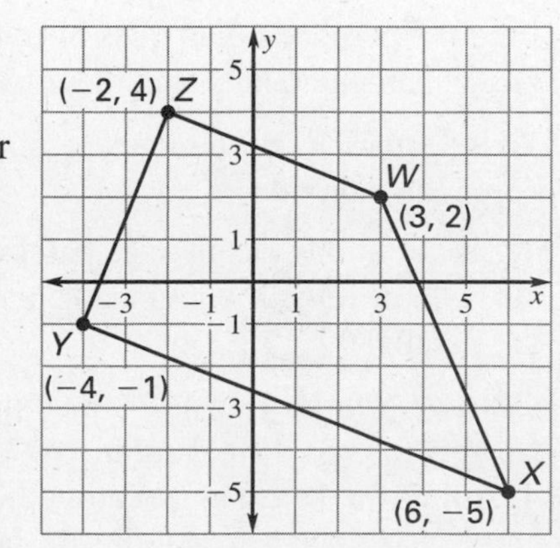

LESSON **5.3**

NAME ______________________________ DATE __________

Practice C

For use with pages 285–291

Write an equation in slope-intercept form of the line shown in the graph.

1.

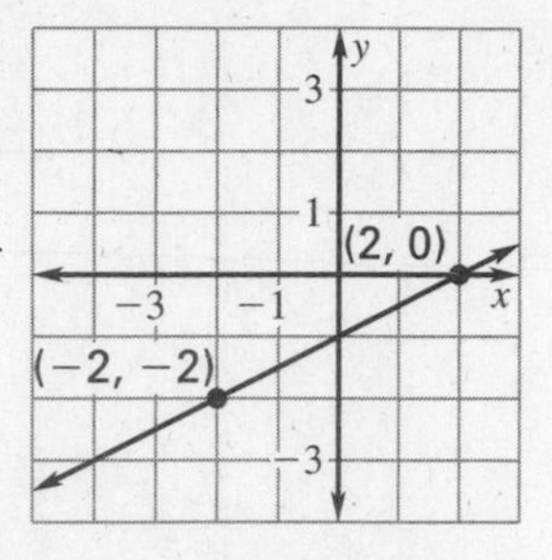

2.

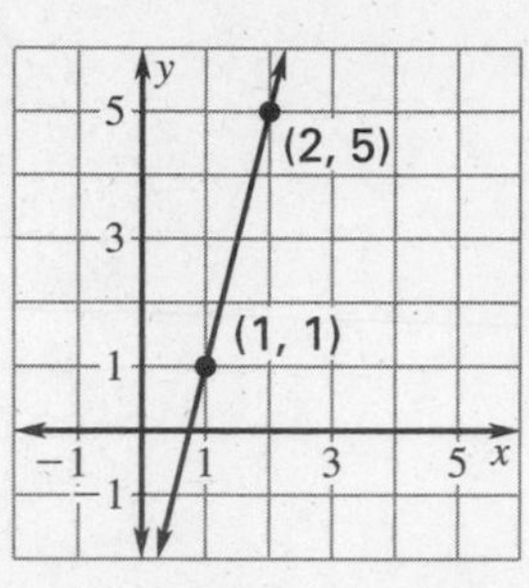

3.

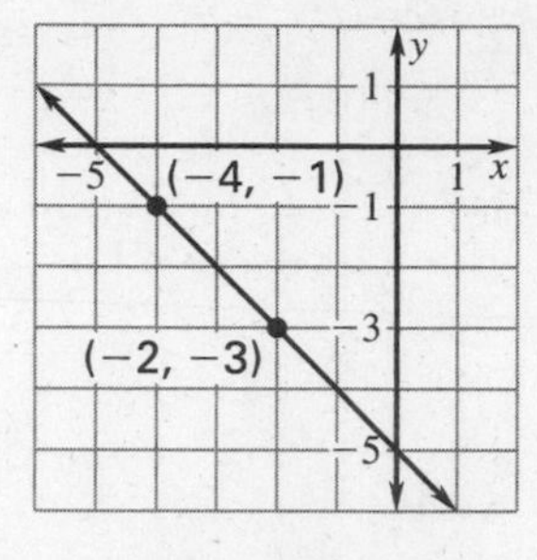

4.

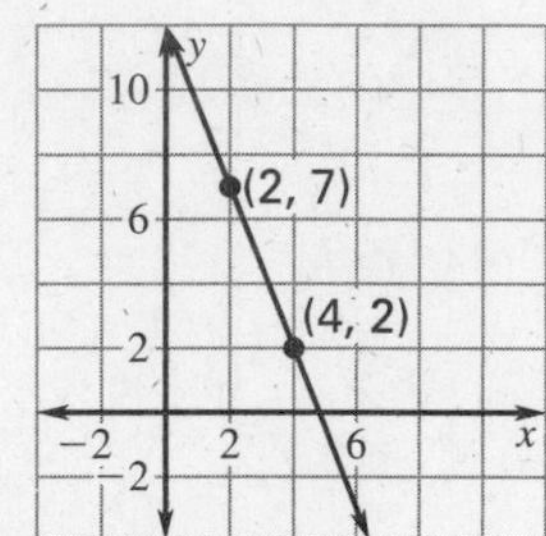

5.

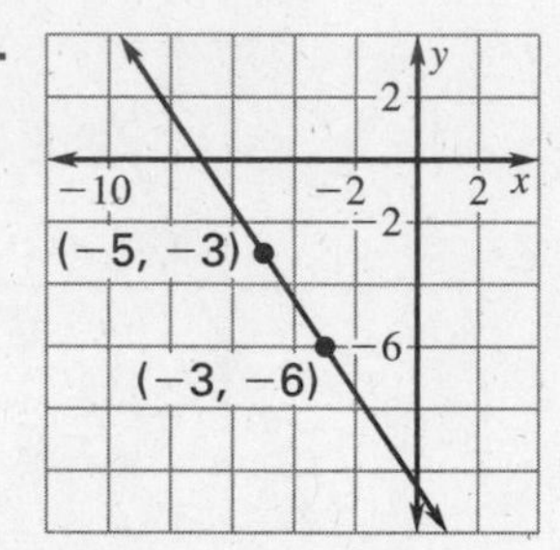

6.

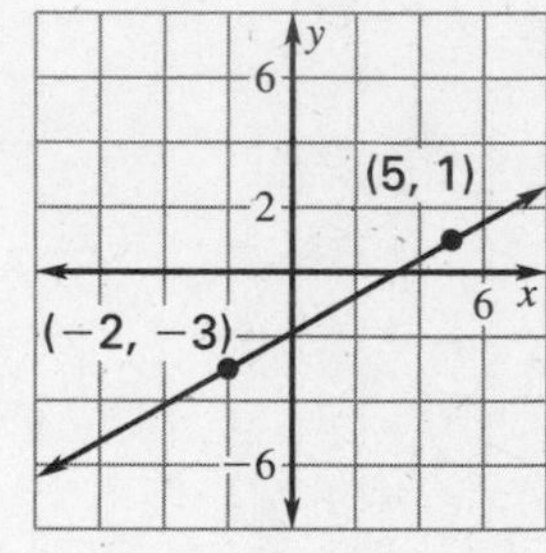

Write an equation in slope-intercept form of the line that passes through the points.

7. $(0, 4), (-11, -62)$

8. $(-1, 3), (6, 38)$

9. $(-3, -5), (6, -2)$

10. $\left(\frac{1}{4}, 10\right), \left(-3, 29\frac{1}{2}\right)$

11. $\left(-1, -1\frac{1}{4}\right), \left(3\frac{1}{2}, -4\frac{5}{8}\right)$

12. $\left(-15, 3\frac{1}{4}\right), \left(\frac{1}{4}, 7\frac{1}{16}\right)$

13. $(3.5, -4.7), (-2.3, -4.7)$

14. $(-1.1, -3.3), (1.4, 4.2)$

15. $(3.26, 5.37), (-1.14, 7.57)$

Give the slope of a line perpendicular to the given line.

16. $y = -7x + 5$

17. $3y = -2x - 4$

18. $3x + 5y = 6$

Geometry Connection **In Exercises 19–21, use the graph.**

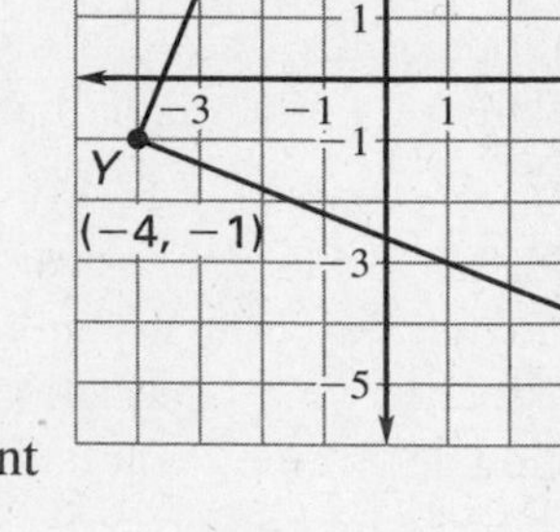

19. Find the perpendicular sides of trapezoid *WXYZ*. How do you know mathematically that these two sides are perpendicular?

20. Write equations of the lines passing through the two perpendicular sides.

21. Write equations of the lines passing through the two parallel sides. How do you know mathematically that these two sides are parallel?

22. *Cassette Singles* In 1994, the number of cassette singles shipped was 81.1 million. In 1997, that number dropped to 42.2 million. Write an equation that gives the number of cassettes shipped, y, in terms of the year, t. Let $t = 4$ represent 1994.

23. Determine the number of cassettes shipped in 1999.

LESSON 5.3

NAME _______________ DATE _______________

Reteaching with Practice

For use with pages 285–291

GOAL **Write an equation of a line given two points on the line and use a linear equation to model a real-life problem**

VOCABULARY

Two different nonvertical lines are **perpendicular** if and only if their slopes are negative reciprocals of each other.

EXAMPLE 1 *Writing an Equation Given Two Points*

Write an equation of the line that passes through the points (1, 5) and (2, 3).

SOLUTION

Find the slope of the line. Let $(x_1, y_1) = (1, 5)$ and $(x_2, y_2) = (2, 3)$.

$m = \dfrac{y_2 - y_1}{x_2 - x_1}$ Write formula for slope.

$= \dfrac{3 - 5}{2 - 1}$ Substitute.

$= \dfrac{-2}{1} = -2$ Simplify.

Find the y-intercept. Let $m = -2$, $x = 1$, and $y = 5$ and solve for b.

$y = mx + b$ Write slope-intercept form.

$5 = (-2)(1) + b$ Substitute -2 for m, 1 for x, and 5 for y.

$5 = -2 + b$ Simplify.

$7 = b$ Solve for b.

Write an equation of the line.

$y = mx + b$ Write slope-intercept form.

$y = -2x + 7$ Substitute -2 for m and 7 for b.

Exercises for Example 1

Write an equation in slope-intercept form of the line that passes through the points.

1. (4, 9) and (1, 6) **2.** (0,7) and (1, −1) **3.** (−2, −3) and (0, 3)

Lesson 5.3

NAME ______________________ DATE __________

Reteaching with Practice

For use with pages 285–291

EXAMPLE 2 Writing Equations of Perpendicular Lines

Write an equation of the line that is perpendicular to the line $y = -3x + 2$ and passes through the point (6, 5).

SOLUTION

The given line has a slope of $m = -3$. A perpendicular line through (6, 5) must have a slope of $m = \frac{1}{3}$. Use this information to find the y-intercept.

$y = mx + b$	Write slope-intercept form.
$5 = \frac{1}{3}(6) + b$	Substitute $\frac{1}{3}$ for m, 6 for x, and 5 for y.
$5 = 2 + b$	Simplify.
$3 = b$	Solve for b.

The y-intercept is $b = 3$.

Write an equation using the slope-intercept form.

$y = mx + b$	Write slope-intercept form.
$y = \frac{1}{3}x + 3$	Substitute $\frac{1}{3}$ for m and 3 for b.

Exercises for Example 2

Write an equation of the line that is perpendicular to the given line and passes through the given point.

4. $y = 2x - 1,\ (2, 4)$ **5.** $y = -\frac{1}{3}x + 2,\ (5, 1)$ **6.** $y = -4x + 5,\ (4, 3)$

NAME ______________________________ DATE __________

Quick Catch-Up for Absent Students

For use with pages 285–291

The items checked below were covered in class on (date missed) __________

Lesson 5.3: Writing Linear Equations Given Two Points

____ **Goal 1:** Write an equation of a line given two points on the line. (pp. 285–286)

Material Covered:

____ Example 1: Writing an Equation Given Two Points

____ Example 2: Writing Equations of Perpendicular Lines

____ **Goal 2:** Use a linear equation to model a real-life problem. (p. 287)

Material Covered:

____ Example 3: Writing and Using a Linear Model

____ Other (specify) ______________________________

____

Homework and Additional Learning Support

____ Textbook (specify) pp. 288–291

____

____ Internet: Extra Examples at www.mcdougallittell.com

____ *Reteaching with Practice* worksheet (specify exercises) __________

____ *Personal Student Tutor* for Lesson 5.3

NAME ______________________ DATE __________

Interdisciplinary Application

For use with pages 285–291

Bald Eagles

BIOLOGY The bald eagle, our national symbol, is making a comeback from the brink of extinction. Although it has been illegal to hunt bald eagles since 1940, when the Bald Eagle Protection Act made it illegal to kill, harm, harass, or possess bald eagles, the eagle was listed as endangered in 1978 in most of the lower 48 states.

The eagle population was decimated after World War II when the pesticide DDT went into widespread use. The pesticide caused the birds to lay thin-shelled eggs that broke during incubation. DDT was banned in the U.S. on December 31, 1972. Since then, the eagle has steadily increased in numbers, and was down listed from endangered to threatened in August 1995. In July 1999, the Fish and Wildlife Service proposed removing it from the threatened list under the Endangered Species Act.

The Raptor Research and Technical Assistance Center coordinates the bald eagle survey throughout the lower 48 states. The survey reported 15,896 eagles in 1994. In 1995, 16,289 eagles were counted.

1. Write a linear equation to model the eagle population. Let x represent the number of years since 1990 and y the eagle population.
2. Graph the equation from Exercise 1.
3. Use the equation from Exercise 1 to estimate the eagle population in the year 2000.
4. In 1996, the number of eagles sighted declined slightly. What factors could have influenced the bald eagle count?
5. If the number of eagles counted in 1999 is 18,362, write a linear equation using year 1999 and year 1995 as your points.
6. Graph the equation from Exercise 5.
7. Use the equation from Exercise 5 to estimate the eagle population in the year 2000.

LESSON 5.3

NAME ______________________________ DATE __________

Challenge: Skills and Applications

For use with pages 285–291

In Exercises 1–3, write an equation in slope-intercept form of the line.

1. through $\left(2\frac{1}{4}, -5\right)$ and $\left(-1\frac{1}{2}, 3\frac{1}{3}\right)$
2. through $\left(-\frac{1}{6}, \frac{2}{3}\right)$ and perpendicular to $4x - 2y = 9$
3. through $(k, -2)$ and perpendicular to $8 - 3x = 9y$

In Exercises 4–7, use the following information.

Rectangle $ABCD$ has vertices at $A(4, 7)$, $B(3, 1)$, and $C(-3, 2)$.

4. Find the equation of the line that contains $\overline{AB}$.
5. Find the equation of the line that contains $\overline{BC}$.
6. Find the equation of the line that contains $\overline{CD}$.
7. Find the equation of the line that contains $\overline{AD}$.

In Exercises 8–11, use the following information.

Suppose that a certain strain of pea plant requires 14 days to reach a height of 6 inches and 30 days to reach a height of 16 inches.

8. Write a linear equation that models the height of the plant after x days.
9. About how many days would it take a plant of this strain to reach a height of 12 inches?
10. What should the height of the plant be after 20 days?
11. According to the model, what should the height of the plant be after zero days? Why do you think this value is negative?

LESSON 5.3

NAME ______________________________ DATE ____________

Quiz 1

For use after Lessons 5.1–5.3

Answers

1. ____________
2. ____________
3. ____________
4. ____________
5. ____________
6. ____________
7. ____________

1. Write an equation of the line whose slope is $m = -3$ and whose y-intercept is $b = 7$. *(Lesson 5.1)*

2. Write an equation of the line shown in the graph. *(Lesson 5.1)*

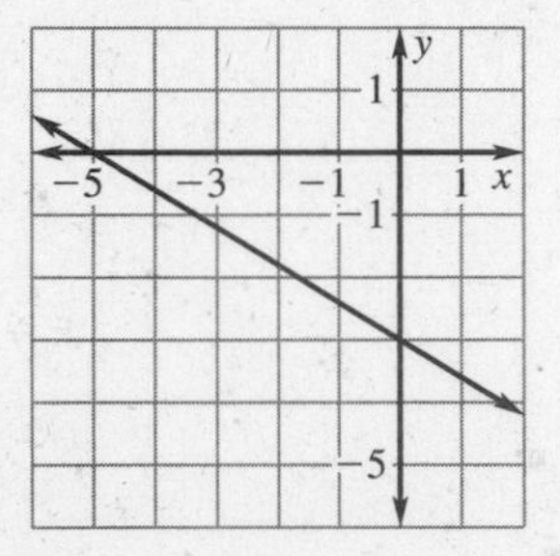

3. Write an equation of the line that passes through $(-3, -9)$ and has a slope of $m = 4$. Write the equation in slope-intercept form. *(Lesson 5.2)*

4. Write an equation of the line shown in the graph. *(Lesson 5.2)*

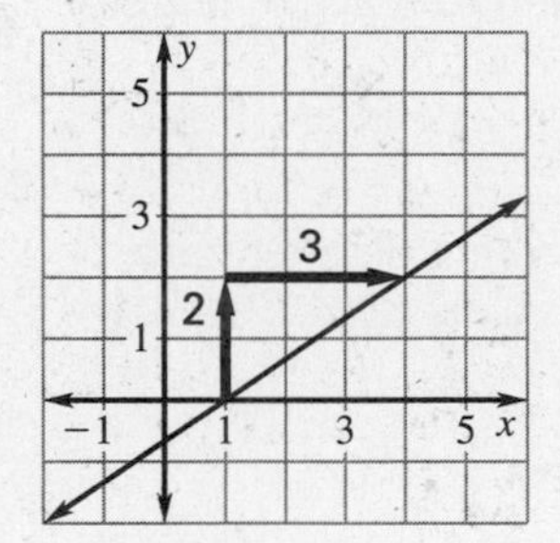

5. Write an equation of the line that is parallel to $y = 2x - 5$ and passes through $(-4, 2)$. *(Lesson 5.2)*

6. Write an equation in slope-intercept form of the line that passes through $(4, 6)$ and $(-8, 3)$. *(Lesson 5.3)*

7. Write an equation of a line that is perpendicular to $y = -5x + 2$ and passes through $(10, 8)$. *(Lesson 5.3)*

Lesson 5.3

LESSON 5.4

TEACHER'S NAME ______________________________ CLASS _______ ROOM ______ DATE ______

Lesson Plan

1-day lesson (See *Pacing the Chapter,* TE pages 270C–270D) **For use with pages 292–299**

1. Find a linear equation that approximates a set of data points.
2. Determine whether there is a positive or negative correlation or no correlation in a set of real-life data.

State/Local Objectives ____

____

✓ Check the items you wish to use for this lesson.

STARTING OPTIONS

____ Homework Check: TE page 288; Answer Transparencies
____ Warm-Up or Daily Homework Quiz: TE pages 292 and 290, CRB page 51, or Transparencies

TEACHING OPTIONS

____ Lesson Opener (Application): CRB page 52 or Transparencies
____ Graphing Calculator Activity with Keystrokes: CRB page 53
____ Examples 1–3: SE pages 293–295
____ Extra Examples: TE pages 293–295 or Transparencies
____ Technology Activity: SE page 299
____ Closure Question: TE page 295
____ Guided Practice Exercises: SE page 296

APPLY/HOMEWORK

Homework Assignment

____ Basic 10–24, 28–31, 35–38, 40, 45, 50
____ Average 10–24, 28–31, 35–38, 40, 45, 50
____ Advanced 10–40, 45, 50

Reteaching the Lesson

____ Practice Masters: CRB pages 54–56 (Level A, Level B, Level C)
____ Reteaching with Practice: CRB pages 57–58 or Practice Workbook with Examples
____ Personal Student Tutor

Extending the Lesson

____ Cooperative Learning Activity: CRB page 60
____ Applications (Real-Life): CRB page 61
____ Challenge: SE page 298; CRB page 62 or Internet

ASSESSMENT OPTIONS

____ Checkpoint Exercises: TE pages 293–295 or Transparencies
____ Daily Homework Quiz (5.4): TE page 298, CRB page 65, or Transparencies
____ Standardized Test Practice: SE page 298; TE page 298; STP Workbook; Transparencies

Notes ____

____

____

TEACHER'S NAME ________________ CLASS ______ ROOM ______ DATE ______

Lesson Plan for Block Scheduling

Half-day lesson (See *Pacing the Chapter,* TE pages 270C–270D) **For use with pages 292–299**

GOALS
1. **Find a linear equation that approximates a set of data points.**
2. **Determine whether there is a positive or negative correlation or no correlation in a set of real-life data.**

State/Local Objectives ________________

CHAPTER PACING GUIDE	
Day	**Lesson**
1	5.1 (all); 5.2 (all)
2	5.3 (all)
3	**5.4 (all)**; 5.5 (begin)
4	5.5 (end); 5.6 (begin)
5	5.6 (end); 5.7 (begin)
6	5.7 (end); Review Ch. 5
7	Assess Ch. 5; 6.1 (all)

✓ Check the items you wish to use for this lesson.

STARTING OPTIONS
____ Homework Check: TE page 288; Answer Transparencies
____ Warm-Up or Daily Homework Quiz: TE pages 292 and 290, CRB page 51, or Transparencies

TEACHING OPTIONS
____ Lesson Opener (Application): CRB page 52 or Transparencies
____ Graphing Calculator Activity with Keystrokes: CRB page 53
____ Examples 1–3: SE pages 293–295
____ Extra Examples: TE pages 293–295 or Transparencies
____ Technology Activity: SE page 299
____ Closure Question: TE page 295
____ Guided Practice Exercises: SE page 296

APPLY/HOMEWORK
Homework Assignment (See also the assignment for Lesson 5.5.)
____ Block Schedule: 10–24, 28–31, 35–38, 40, 45, 50

Reteaching the Lesson
____ Practice Masters: CRB pages 54–56 (Level A, Level B, Level C)
____ Reteaching with Practice: CRB pages 57–58 or Practice Workbook with Examples
____ Personal Student Tutor

Extending the Lesson
____ Cooperative Learning Activity: CRB page 60
____ Applications (Real-Life): CRB page 61
____ Challenge: SE page 298; CRB page 62 or Internet

ASSESSMENT OPTIONS
____ Checkpoint Exercises: TE pages 293–295 or Transparencies
____ Daily Homework Quiz (5.4): TE page 298, CRB page 65, or Transparencies
____ Standardized Test Practice: SE page 298; TE page 298; STP Workbook; Transparencies

Notes ________________

NAME ______________________ DATE __________

Available as a transparency

Warm-Up Exercises

For use before Lesson 5.4, pages 292–299

Write an equation in slope-intercept form of the line that passes through the points.

1. (5, 32), (7, 16)

2. (0, 160), (25, 610)

3. (−12, −15), (−18, −12)

Daily Homework Quiz

For use after Lesson 5.3, pages 285–291

Write an equation in slope-intercept form of the line that passes through the points.

1. a. (−3, 1), (2, −2)

b. (2, 3), (−4, 0)

2. Write an equation of the line through (2, −1) that is perpendicular to $y = \frac{1}{3}x + 1$.

3. Show that $\triangle ABC$ is a right triangle.

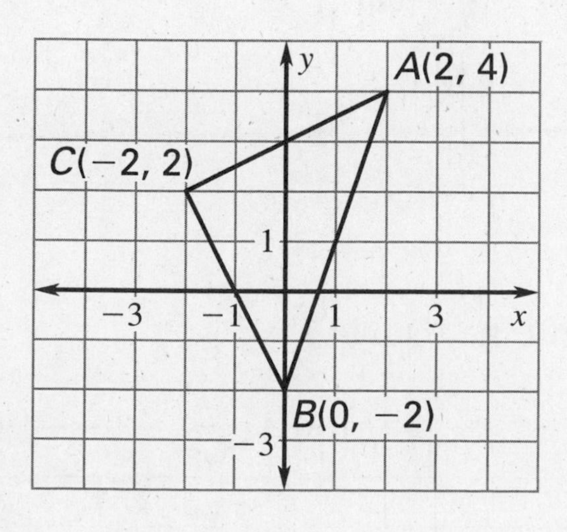

LESSON 5.4

NAME ______________________ DATE __________

Available as a transparency

Application Lesson Opener

For use with pages 292–298

The data in the graph show the resting heart rates and ages of 24 students in an aerobics class.

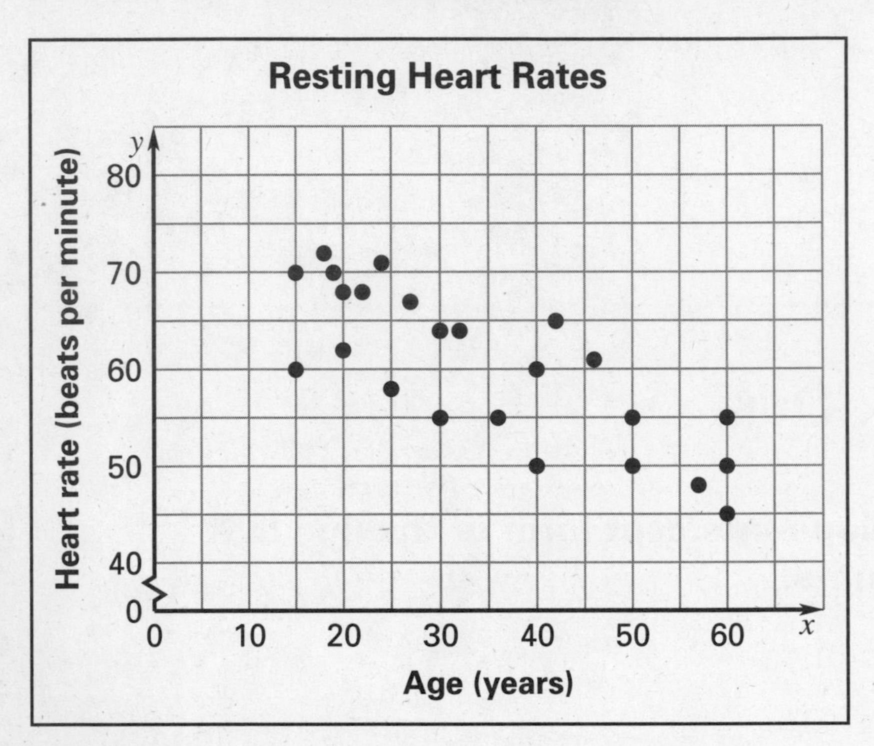

1. Graph the equations on the graph above.

a. $y = -\frac{2}{5}x + 80$

b. $y = -\frac{2}{5}x + 75$

c. $y = 60$

d. $y = -\frac{3}{5}x + 75$

2. Which equation seems to best represent the data? Why?

Lesson 5.4

NAME ______________________ DATE ____________

Graphing Calculator Activity Keystrokes

For use with Technology Activity 5.4 on page 299

TI-82

STAT 1

Enter *x*-values in L1.

38 ENTER 28 ENTER 56 ENTER 56 ENTER

24 ENTER 77 ENTER 40 ENTER 46 ENTER

Enter *y*-values in L2.

62 ENTER 46 ENTER 102 ENTER 88 ENTER

36 ENTER 113 ENTER 69 ENTER 60 ENTER

2nd [STAT PLOT] 1

Choose the following.

On; Type ⋰ ; Xlist: L1; Ylist: L2; Mark ▫

WINDOW ENTER 0 ENTER 80 ENTER

10 ENTER 0 ENTER 120 ENTER 10 ENTER

GRAPH

STAT ► 5 2nd [L1] , 2nd [L2] ENTER

Y= VARS 5 ► ► 7

GRAPH

TI-83

STAT 1

Enter *x*-values in L1.

38 ENTER 28 ENTER 56 ENTER 56 ENTER

24 ENTER 77 ENTER 40 ENTER 46 ENTER

Enter *y*-values in L2.

62 ENTER 46 ENTER 102 ENTER 88 ENTER

36 ENTER 113 ENTER 69 ENTER 60 ENTER

2nd [STAT PLOT] 1

Choose the following.

On; Type ⋰ ; Xlist: L1; Ylist: L2; Mark ▫

WINDOW 0 ENTER 80 ENTER 10 ENTER

0 ENTER 120 ENTER 10 ENTER

GRAPH

STAT ► 4 2nd [L1] , 2nd [L2] ENTER

Y= VARS 5 ► ► 1

GRAPH

SHARP EL-9600c

STAT [A] ENTER

Enter *x*-values in L1.

38 ENTER 28 ENTER 56 ENTER 56 ENTER

24 ENTER 77 ENTER 40 ENTER 46 ENTER

Enter *y*-values in L2.

62 ENTER 46 ENTER 102 ENTER 88 ENTER

36 ENTER 113 ENTER 69 ENTER 60 ENTER

2nd [STAT PLOT] [A] ENTER

Choose the following.

On; Data XY; ListX: L1; ListY: L2

2ndF [STAT PLOT] [G] 3

WINDOW 0 ENTER 80 ENTER 10 ENTER

0 ENTER 120 ENTER 10 ENTER

GRAPH 2ndF [Quit]

STAT [D] 0 2

(2ndF [L1] , 2ndF [L2] , VARS [A]

ENTER [A] 1) ENTER GRAPH

CASIO CFX-9850GA PLUS

From the main menu, choose STAT.

Enter *x*-values in List 1.

38 EXE 28 EXE 56 EXE 56 EXE 24 EXE 77 EXE

40 EXE 46 EXE

Enter *y*-values in List 2.

62 EXE 46 EXE 102 EXE 88 EXE 36 EXE 113 EXE

69 EXE 60 EXE

F1 F6

Choose the following.

Graph Type: Scatter; XList: List 1; YList: List 2;
Frequency: 1; Mark Type: ▫

EXIT

SHIFT F3 0 EXE 80 EXE 10 EXE 0 EXE 120 EXE

10 EXE EXIT

F1 F1 F1 F5 EXE F6

LESSON

5.4

NAME ______________________ DATE ________

Practice A

For use with pages 292–298

Decide whether x and y suggest a linear relationship.

1. 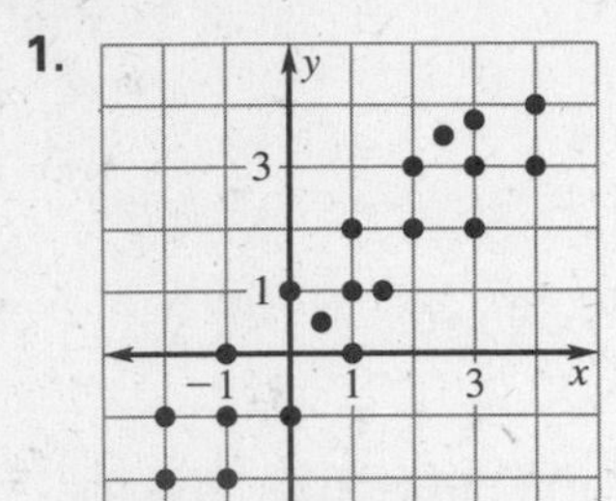

2. 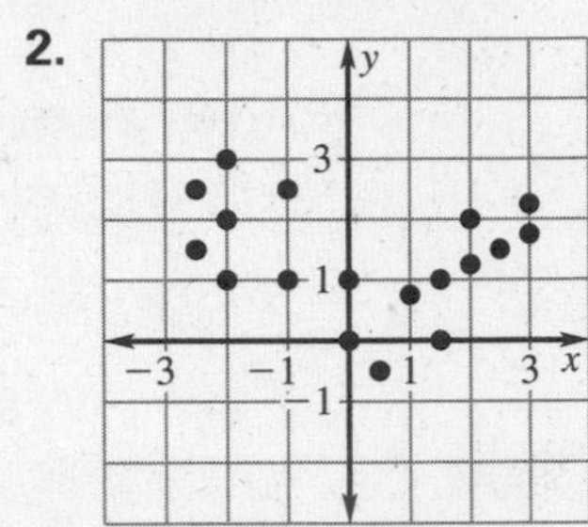

3. 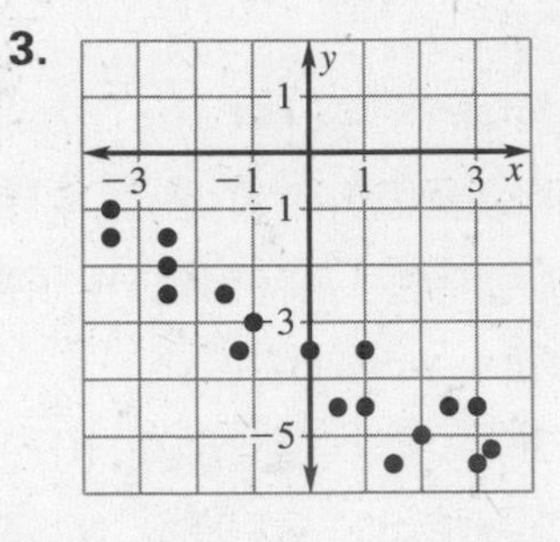

Copy the graph and draw a best-fitting line for the scatter plot. Write an equation of your line.

4. 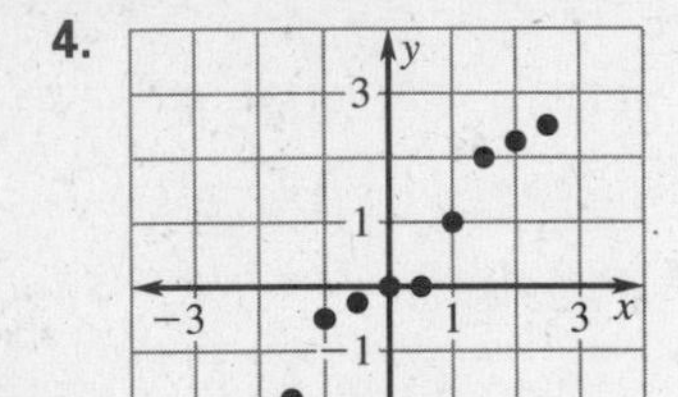

5. 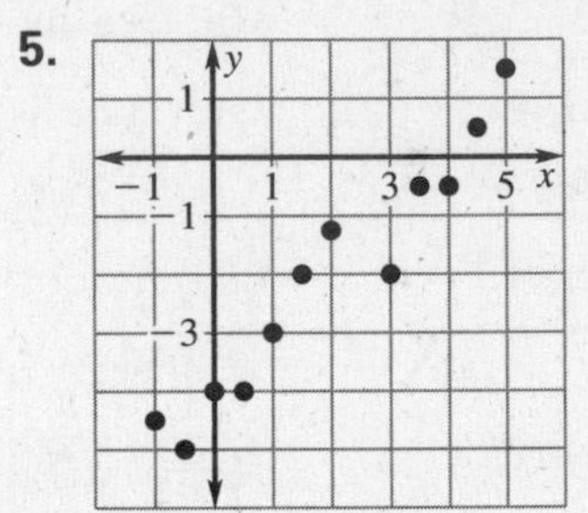

6. 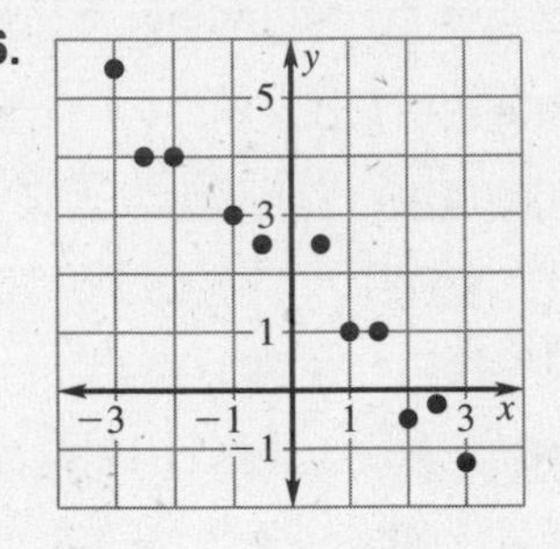

Draw a scatter plot of the data. Draw a best-fitting line and write an equation of the line.

7.

x	1	2	3	4	5	6
y	2	5	5	8	11	12

8.

x	1	2	3	4	5	6
y	4	6	6	7	7	9

9.

x	0	1.1	1.9	2.5	3.1	4.3
y	0.8	2.2	2.9	3.6	4.0	5.3

10.

x	3.1	3.8	4.5	6.0	6.3	7.1
y	1.0	1.7	2.5	4.1	4.4	5.0

11. ***Weight Loss*** The scatter plot below shows the weight loss per week of a dieter. In the graph, y represents the person's weight in pounds and x represents the weeks of the diet. Find an equation of the line that you think best fits this data. Then use the equation to find the dieter's approximate weight after 10 weeks.

Weight (pounds)

Time (weeks)

12. ***Milk Consumption*** The table below shows the average number of gallons of milk a family drinks per week. Sketch a scatter plot for this data and find an equation, and use it to find the milk consumption in one week of a 7-member family.

Family Size	*Number of Gallons of Milk*
1	1
2	1.5
3	2.2
4	3.8
5	4.7
6	5

LESSON 5.4

NAME ______________________________ DATE ____________

Practice B

For use with pages 292–298

Draw a scatter plot of data that have the given correlation.

1. Positive
2. Negative
3. None

Copy the graph and draw a best-fitting line for the scatter plot. Write an equation of your line.

4.

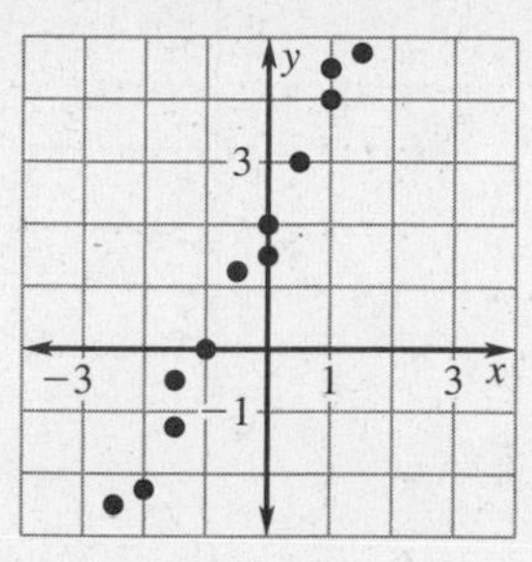

5.

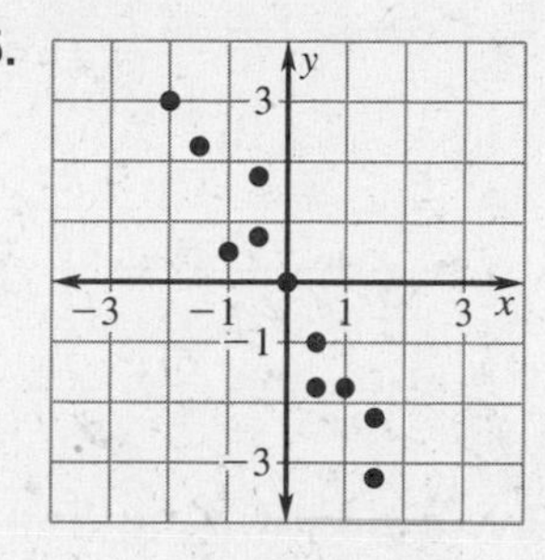

6.

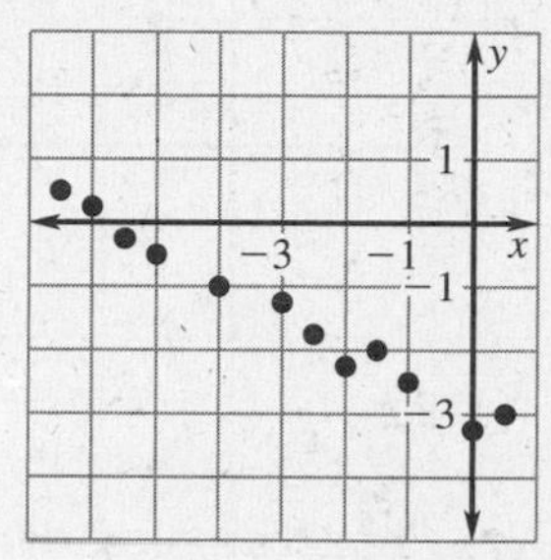

Draw a scatter plot of the data. Draw a best-fitting line and write an equation of the line.

7.

x	y
0	-4.8
1.6	-3.6
2.4	-2.7
3.1	-1.8
3.9	-1.3
4.7	-0.4

8.

x	y
3.5	-5.7
4.1	-7.5
5.6	-10
6.8	-12.4
7.3	-13.8
8.9	-16.6

9.

x	y
1.1	5.0
1.8	4.1
2.0	4.1
2.6	3.5
3.2	2.8
4.0	2.1

10.

x	y
0.8	4.6
1.5	5.8
1.9	7.0
2.4	7.6
2.8	8.6
3.5	10.1

Weight Loss **In Exercises 11 and 12, use the following information.**

The graph below shows the weight loss per week of a dieter. In the graph, y represents the person's weight in pounds and x represents the weeks of the diet.

11. Find an equation of the line that you think best fits this data. Then use the equation to find the dieter's approximate weight after 10 weeks.

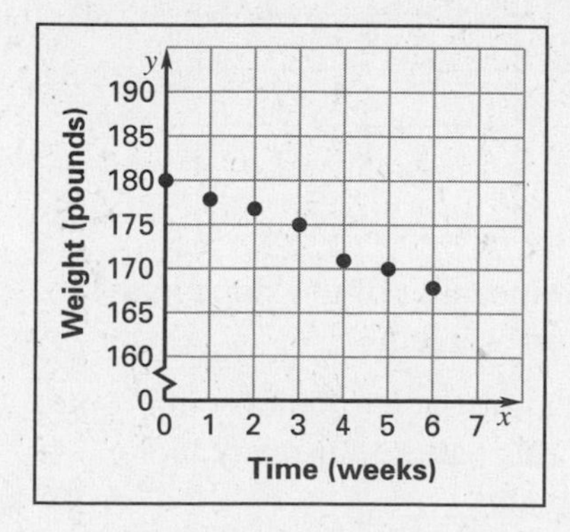

12. Do you think this graph could continue in this pattern for a year? Explain.

Milk Consumption **In Exercises 13 and 14, use the following information.**

The table below shows the average number of gallons of milk a family drinks per week.

13. Sketch a scatter plot for this data and find an equation, and use it to find the milk consumption in one week of a 7-member family.

Family Size	*Number of Gallons of Milk*
1	1
2	1.5
3	2.2
4	3.8
5	4.7
6	5

14. Do you think this table of data could continue in this pattern for many more people? Explain.

Lesson 5.4

LESSON
5.4

NAME _______________ DATE _______

Practice C

For use with pages 292–298

Copy the graph and draw a best-fitting line for the scatter plot. Write an equation of your line.

1.

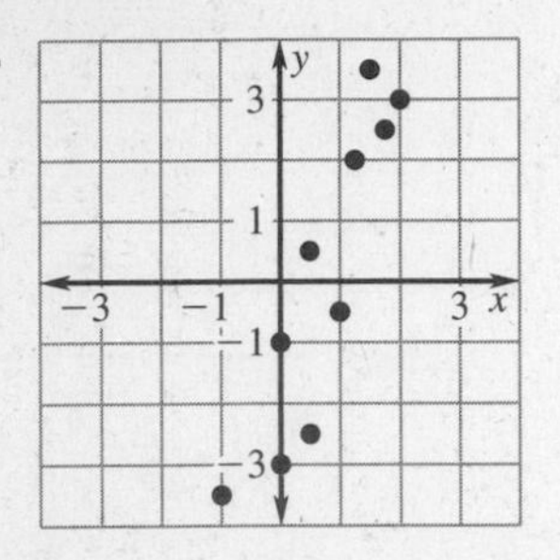

2.

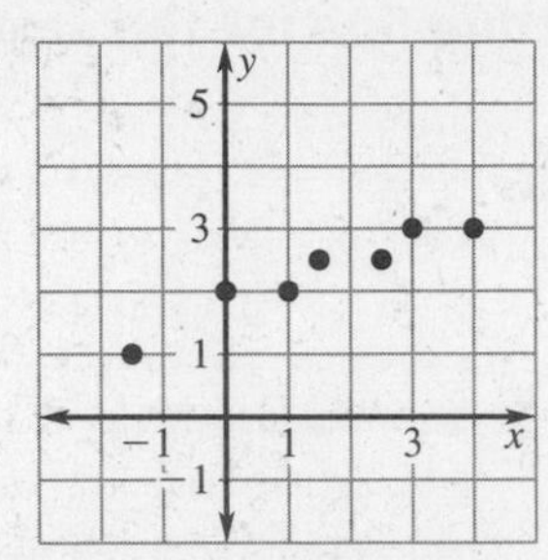

3.

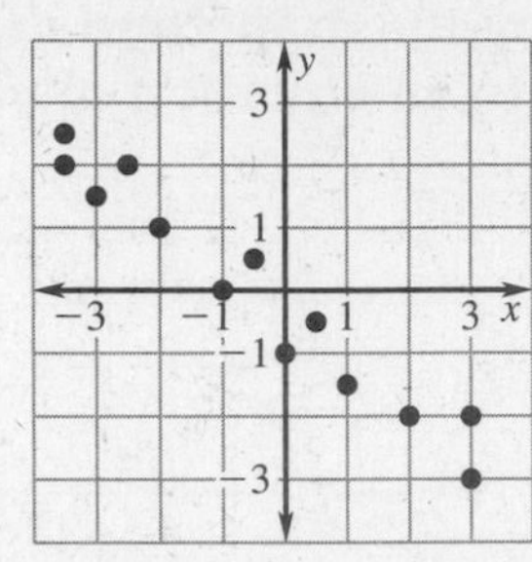

Draw a scatter plot of the data. Draw a best-fitting line and write an equation of the line.

4.

x	y
0	0.9
2.4	3.4
3.7	4.5
4.1	5.3
6.3	7.5
7.4	8.3

5.

x	y
−3.2	−9.5
−2.5	−8.2
−1.8	−6.5
0.5	−1.9
1.3	−0.6
2.4	1.7

6.

x	y
−1.7	4.6
2.5	2.9
5.3	1.4
8.4	−0.4
10.8	−1.2
11.9	−2

7.

x	y
3.2	−13.3
3.9	−15.6
4.7	−18.8
5.5	−22.4
6.8	−27.8
7.3	−29.3

8. ***Weight Loss*** The scatter plot below shows the weight loss per week of a dieter. In the graph, y represents the person's weight in pounds and x represents the weeks of the diet. Find an equation of the line that you think best fits this data. Then use the equation to find the dieter's approximate weight after 10 weeks.

y
190
185
180
175
170
165
160
0
0 1 2 3 4 5 6 7 x
Weight (pounds)
Time (weeks)

9. Interpret the meaning of the slope in the context of the problem.

10. Do you think this graph could continue in this pattern for a year? Explain.

11. ***Milk Consumption*** The table below shows the average number of gallons of milk a family drinks per week. Sketch a scatter plot for this data and find an equation, and use it to find the milk consumption in one week of a 7-member family.

Family Size	*Number of Gallons of Milk*
1	1
2	1.5
3	2.2
4	3.8
5	4.7
6	5

12. Interpret the meaning of the slope in the context of the problem.

13. Do you think this table of data could continue in this pattern for many more people? Explain.

NAME ______________________________ DATE ____________

Reteaching with Practice

For use with pages 292–298

GOAL **Find a linear equation that approximates a set of data points and determine a correlation in a set of real-life data**

VOCABULARY

The line that best fits all of the data points is called the **best-fitting line.**

Positive correlation means that the points can be approximated by a line with a positive slope.

Negative correlation means that the points can be approximated by a line with a negative slope.

Points that cannot be approximated by a line have **relatively no correlation.**

EXAMPLE 1 *Approximating a Best-Fitting Line*

Draw a scatter plot for the data. If possible, draw a line that corresponds closely to the data and write an equation of the line.

x	1	2	3	4	5	6
y	3	5	8	9	11	12

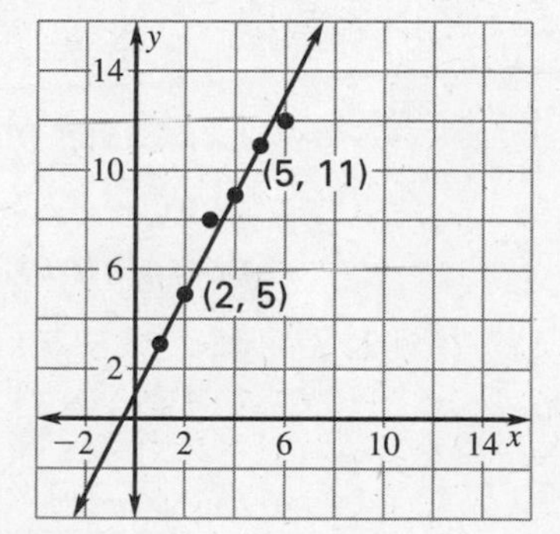

SOLUTION

Plot the points given by the ordered pairs (x, y). Sketch the line that appears to closely fit the points.

Next, find two points that lie on the line. From the graph, choose the points $(2, 5)$ and $(5, 11)$. Calculate the slope of the line through these two points.

$m = \dfrac{y_2 - y_1}{x_2 - x_1}$ Write slope formula.

$m = \dfrac{11 - 5}{5 - 2}$ Substitute.

$m = 2$ Simplify.

To find the y-intercept of the line, use the values $m = 2$, $x = 2$, and $y = 5$ in the slope-intercept form.

$y = mx + b$ Write slope-intercept form.

$5 = (2)(2) + b$ Substitute 2 for m, 2 for x, and 5 for y.

$5 = 4 + b$ Simplify.

$1 = b$ Solve for b.

An approximate equation of the line is $y = 2x + 1$.

NAME ______________________ DATE __________

Reteaching with Practice

For use with pages 292–298

Exercises for Example 1

In Exercises 1 and 2, draw a scatter plot of the data. If possible, draw a line that corresponds closely to the data and write an equation of the line.

1.

x	1	2	3	4	5	6
y	7	0	1	0	7	6

2.

x	1	2	3	4	5	6
y	1	0	-2	-2	-3	-4

EXAMPLE 2 *Determining the Correlation of x and y*

State whether x and y have a *positive correlation,* a *negative correlation,* or *relatively no correlation.*

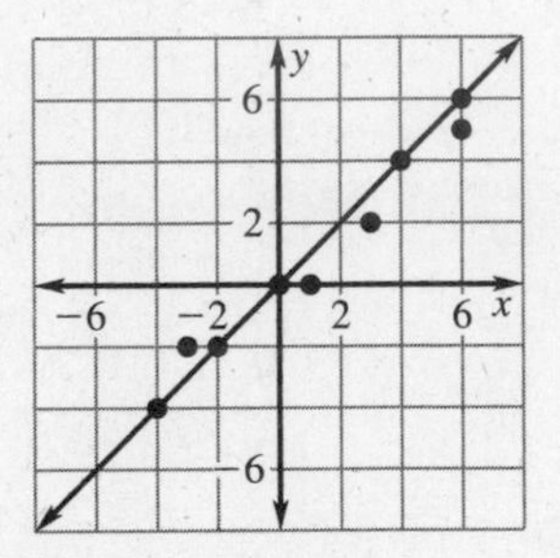

Solution

Because the points can be approximated by a line with positive slope, x and y have a positive correlation.

Exercises for Example 2

In Exercises 3 and 4, state whether x and y have a *positive correlation,* a *negative correlation,* or *relatively no correlation.*

3.

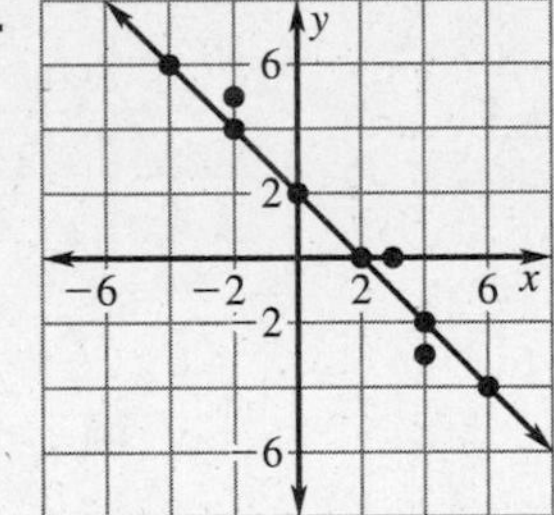

4.

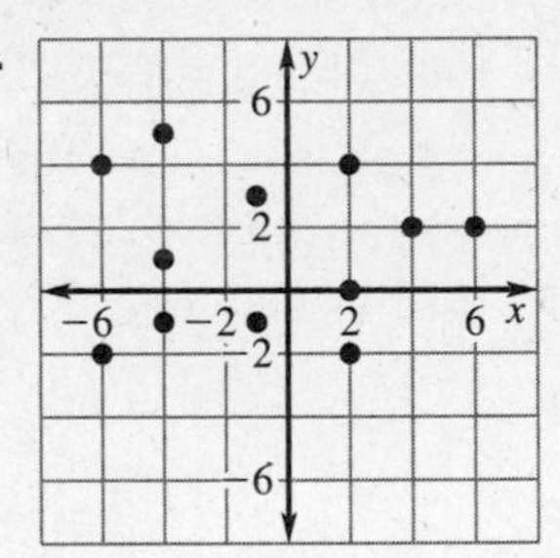

LESSON 5.4

NAME ______________________________ DATE __________

Quick Catch-Up for Absent Students

For use with pages 292–299

The items checked below were covered in class on (date missed) ___________

Lesson 5.4: Fitting a Line to Data

____ **Goal 1:** Find a linear equation that approximates a set of data points. (pp. 292–293)

Material Covered:

____ Activity: Approximating a Best-Fitting Line

____ Study Help: Study Tip

____ Example 1: Approximating a Best-Fitting Line

____ Example 2: Approximating a Best-Fitting Line

Vocabulary:

best-fitting line, p. 292

____ **Goal 2:** Determine whether there is a positive or negative correlation or no correlation in a set of real-life data. (p. 295)

Material Covered:

____ Example 3: Interpreting Correlation

Vocabulary:

positive correlation, p. 295
negative correlation, p. 295
no correlation, p. 295

Activity 5.4: Best-Fitting Lines (p. 299)

____ **Goal:** Find the best-fitting line of a set of data using a graphing calculator.

____ Study Help: Keystroke Help

____ Study Help: Look Back

____ Other (specify) ______________________________

____

Homework and Additional Learning Support

____ Textbook (specify) pp. 296–298

____

____ *Reteaching with Practice* worksheet (specify exercises) ______________

____ *Personal Student Tutor* for Lesson 5.4

NAME ______________________ DATE __________

Cooperative Learning Activity

For use with pages 292–298

GOAL **To make and analyze a scatter plot of data and determine whether there is a positive or negative correlation, or no correlation**

Materials: graph paper, pencil, ruler or meter stick

Exploring Scatter Plots and Correlation

In this activity, you and your partner will collect data from students in your class. After plotting the data on graph paper, you will draw lines of best fit and calculate the equations of those lines.

Instructions

1. Measure the height of ten students in your class and the length of their forearm. Also record the age of these students in months, and their most recent math test score.
2. Graph the data, two sets at a time. You will have six scatter plots when you are finished.
3. Draw the line of best fit for each scatter plot and write the equation of each line.
4. For each scatter plot state whether the data have a *positive correlation*, a *negative correlation*, or *no correlation.*

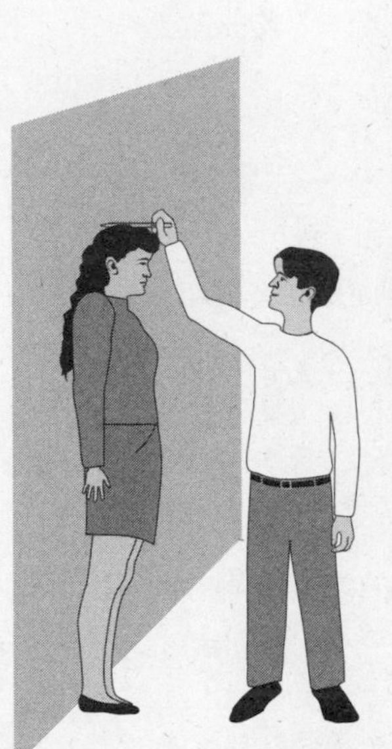

Analyzing the Results

1. Which sets of data have positive correlation? Which have negative correlation? Which appear to have no correlation?

2. What other types of data could you collect that would have positive, negative, or no correlation? Think of at least one example of each.

NAME ______________________________ DATE __________

Real-Life Application: When Will I Ever Use This?

For use with pages 292–298

Cellular Phone Use

The invention of the cellular phone has changed the way people communicate. According to figures published by the cellular telecommunications industry association (CTIA), the cell phone industry has blossomed into a revenue maker of over thirty billion dollars a year, and accounts for 1.3 million jobs in the U.S.

The term cellular is used because each geographic region of coverage is broken up into cells of coverage. Each cell has a radio transmitter and control equipment. The original cell phone worked on an analog signal, however newer technology relies on digital technology.

Cell phone use has become so prevalent that many people now see them as intrusive. This had led many schools and businesses to make rules restricting the time and place where cell phones may be used.

The CTIA survey below lists estimated subscribers in thousands.

Year	*Subscribers (in Thousands)*
1990	5,283
1991	7,557
1992	11,033
1993	16,009
1994	24,134
1995	33,786
1996	44,043
1997	55,312
1998	69,209

1. Make a scatter plot of the data. Let x represent the number of years since 1990.
2. Find an equation that best fits the data in Exercise 1.
3. Approximate the number of cell phone users in 2000.
4. Make a scatter plot of the data from 1993 till 1998. Let x represent the number of years since 1993.
5. Find an equation that best fits the data in Exercise 4.
6. Approximate the number of cell phone users in 2000 using the new equation.

NAME ______________________ DATE __________

Challenge: Skills and Applications

For use with pages 292–298

In Exercises 1–4, state whether the two variables have a *positive correlation*, a *negative correlation*, or *no correlation*. Explain.

1. an individual teacher's years of experience and salary
2. shoe size and scores on college admission tests
3. weeks on a successful weight-loss plan and a person's weight
4. time spent training and distance a person can run

In Exercises 5–6, use the information in the table.

Year born	1920	1930	1940	1950	1960	1970	1980	1990
Expected years of life	54.1	59.7	62.9	68.2	69.7	70.8	73.7	75.4

5. Let x equal the number of years since 1900. Use a scatter plot to find an equation of the line that you think best fits the data.
6. Use the equation from Exercise 5 to estimate the expected years of life for someone born in the year 2000.

In Exercises 7–11, use the table which shows the U.S. labor force, excluding agricultural, self-employed, and unemployed workers.

Year	1930	1940	1950	1960	1970	1980	1990
Workers (millions)	29.4	32.3	45.2	54.2	70.9	90.6	103.9

7. Let x equal the number of years since 1900. Use the scatter plot to find an equation of the line that you think best fits the data.
8. How many pairs of points can be formed from the 7 data points in the table? For example, one pair consists of the 1930 and 1970 data points and another pair consists of the 1960 and 1970 data points.
9. A line can be drawn through any pair of points. Find the slope of the line through each pair of points you counted in Exercise 8. Round to the nearest hundredth.
10. Find the mean of all the slopes you listed in Exercise 9.
11. How does the mean you found in Exercise 10 compare to the slope of the line of best fit you found in Exercise 7?

LESSON 5.5

TEACHER'S NAME ______________________ CLASS ______ ROOM ______ DATE ______

Lesson Plan

2-day lesson (See *Pacing the Chapter,* TE pages 270C–270D) For use with pages 300–306

GOALS
1. **Use the point-slope form to write an equation of a line.**
2. **Use the point-slope form to model a real-life situation.**

State/Local Objectives ____

____

✓ Check the items you wish to use for this lesson.

STARTING OPTIONS

____ Homework Check: TE page 296; Answer Transparencies
____ Warm-Up or Daily Homework Quiz: TE pages 300 and 298, CRB page 65, or Transparencies

TEACHING OPTIONS

____ Motivating the Lesson: TE page 301
____ Lesson Opener (Activity): CRB page 66 or Transparencies
____ Examples: Day 1: 1–2, SE pages 300–301; Day 2: 3, SE page 302
____ Extra Examples: Day 1: TE page 301 or Transp.; Day 2: TE page 302 or Transp.; Internet
____ Closure Question: TE page 302
____ Guided Practice: SE page 303; Day 1: Exs. 1–17; Day 2: none

APPLY/HOMEWORK

Homework Assignment

____ Basic Day 1: 18–20, 22–46 even; Day 2: 48–56 even, 57–59, 64–66, 75, 80, 85–87; Quiz 2: 1–12
____ Average Day 1: 18–20, 22–46 even; Day 2: 48–56 even, 57–61, 64–66, 75, 80, 85–87; Quiz 2: 1–12
____ Advanced Day 1: 18–20, 22–46 even; Day 2: 48–56 even, 57–70, 75, 80, 85–87; Quiz 2: 1–12

Reteaching the Lesson

____ Practice Masters: CRB pages 67–69 (Level A, Level B, Level C)
____ Reteaching with Practice: CRB pages 70–71 or Practice Workbook with Examples
____ Personal Student Tutor

Extending the Lesson

____ Applications (Interdisciplinary): CRB page 73
____ Math & History: SE page 306; CRB page 74; Internet
____ Challenge: SE page 305; CRB page 75 or Internet

ASSESSMENT OPTIONS

____ Checkpoint Exercises: Day 1: TE page 301 or Transp.; Day 2: TE page 302 or Transp.
____ Daily Homework Quiz (5.5): TE page 305, CRB page 79, or Transparencies
____ Standardized Test Practice: SE page 305; TE page 305; STP Workbook; Transparencies
____ Quiz (5.4–5.5): SE page 306; CRB page 76

Notes ____

____

____

LESSON 5.5

TEACHER'S NAME ______________ CLASS ______ ROOM ______ DATE ______

Lesson Plan for Block Scheduling

1-day lesson (See *Pacing the Chapter,* TE pages 270C–270D) For use with pages 300–306

GOALS
1. Use the point-slope form to write an equation of a line.
2. Use the point-slope form to model a real-life situation.

State/Local Objectives ______________

CHAPTER PACING GUIDE	
Day	**Lesson**
1	5.1 (all); 5.2 (all)
2	5.3 (all)
3	5.4 (all); **5.5 (begin)**
4	**5.5 (end)**; 5.6 (begin)
5	5.6 (end); 5.7 (begin)
6	5.7 (end); Review Ch. 5
7	Assess Ch. 5; 6.1 (all)

✓ Check the items you wish to use for this lesson.

STARTING OPTIONS

____ Homework Check: TE page 296; Answer Transparencies
____ Warm-Up or Daily Homework Quiz: TE pages 300 and 298, CRB page 65, or Transparencies

TEACHING OPTIONS

____ Motivating the Lesson: TE page 301
____ Lesson Opener (Activity): CRB page 66 or Transparencies
____ Examples: Day 3: 1–2, SE pages 300–301; Day 4: 3, SE page 302
____ Extra Examples: Day 3: TE page 301 or Transp.; Day 4: TE page 302 or Transp.; Internet
____ Closure Question: TE page 302
____ Guided Practice: SE page 303; Day 3: Exs. 1–17; Day 4: none

APPLY/HOMEWORK

Homework Assignment (See also the assignments for Lessons 5.4 and 5.6.)

____ Block Schedule: Day 3: 18–20, 22–46 even; Day 4: 48–56 even, 57–61, 64–66, 75, 80, 85–87; Quiz 2: 1–12

Reteaching the Lesson

____ Practice Masters: CRB pages 67–69 (Level A, Level B, Level C)
____ Reteaching with Practice: CRB pages 70–71 or Practice Workbook with Examples
____ Personal Student Tutor

Extending the Lesson

____ Applications (Interdisciplinary): CRB page 73
____ Math & History: SE page 306; CRB page 74; Internet
____ Challenge: SE page 305; CRB page 75 or Internet

ASSESSMENT OPTIONS

____ Checkpoint Exercises: Day 3: TE page 301 or Transp.; Day 4: TE page 302 or Transp.
____ Daily Homework Quiz (5.5): TE page 305, CRB page 79, or Transparencies
____ Standardized Test Practice: SE page 305; TE page 305; STP Workbook; Transparencies
____ Quiz (5.4–5.5): SE page 306; CRB page 76

Notes ______________

NAME ______________________ DATE ____________

Available as a transparency

Warm-Up Exercises

For use before Lesson 5.5, pages 300–306

Find the slope of the line through the points.

1. $(-2, 3), (5, -1)$

2. $(6, -2), (-1, -5)$

Solve each equation for *y*.

3. $y - 21 = \frac{3}{5}(x + 10)$

4. $y + 1 = -2(x - 3)$

Daily Homework Quiz

For use after Lesson 5.4, pages 292–299

Use the table for Exercises 1–3.

x	1	2	3	4	5
y	10	12	15	17	20

1. a. Draw a scatter plot of the data.

b. Draw a line that corresponds closely to the data.

2. Write an equation of the line.

3. Do x and y have a *positive correlation,* a *negative correlation,* or *relatively no correlation*?

LESSON 5.5

NAME ____________________ DATE ________

Available as a transparency

Activity Lesson Opener

For use with pages 300–306

SET UP: Work with a partner.

YOU WILL NEED: • straightedge • graph paper

1. a. Graph the line that passes through the point $(1, 4)$ and has a slope of 2.

b. On the same coordinate plane, graph the equation $y - 4 = 2(x - 1)$.

c. What is true about the two lines?

d. How is the ordered pair related to the equation in part (b)?

e. How is the slope related to the equation in part (b)?

2. a. Graph the line that passes through the point $(-2, 3)$ and has a slope of -3.

b. On the same coordinate plane, graph the equation $y - 3 = -3(x + 2)$.

c. What is true about the two lines?

d. How is the ordered pair related to the equation in part (b)?

e. How is the slope related to the equation in part (b)?

3. a. Graph the line that passes through the point $(-3, -5)$ and has a slope of $\frac{2}{3}$.

b. On the same coordinate plane, graph the equation $y + 5 = \frac{2}{3}(x + 3)$.

c. What is true about the two lines?

d. How is the ordered pair related to the equation in part (b)?

e. How is the slope related to the equation in part (b)?

NAME ______________________ DATE __________

Practice A

For use with pages 300–306

Find the slope of the line passing through the given points.

1. (2, 4), (3, 10) **2.** (0, 5), (−2, 3) **3.** (2, 5), (−2, −3)

4. (−4, −2), (−1, 4) **5.** (7, −1), (−1, 3) **6.** (3, −2), (−5, −2)

Write an equation in point-slope form of the line.

7.

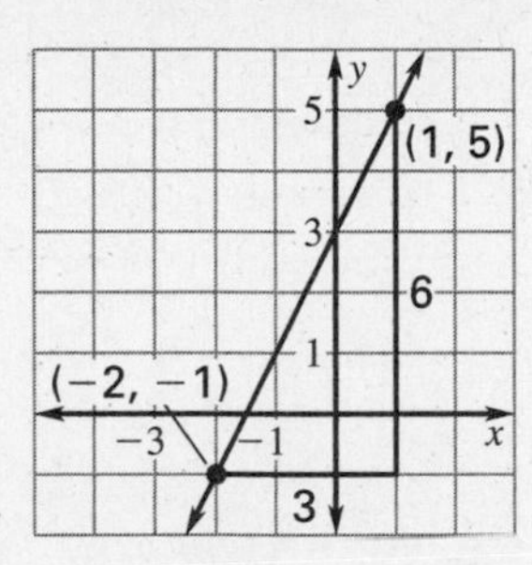

8.

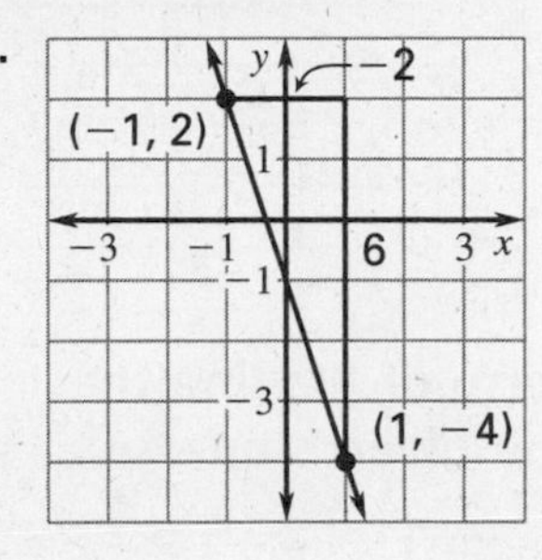

9.

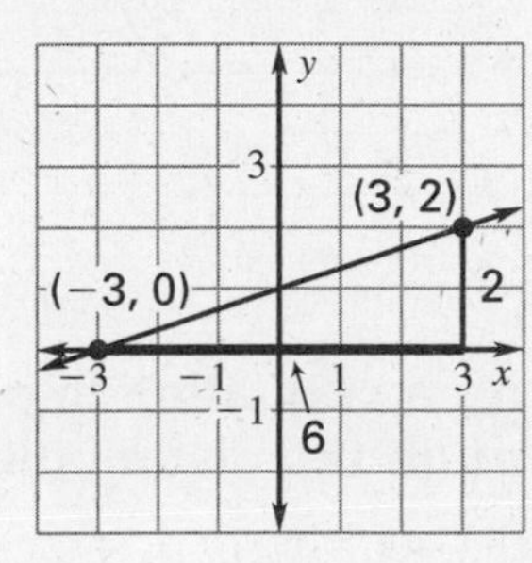

Write an equation in point-slope form of the line that passes through the given point and has the given slope.

10. (2, 5), $m = 3$ **11.** (1, 4), $m = 2$ **12.** (−2, 0), $m = \frac{1}{2}$

13. (3, 7), $m = 1$ **14.** (−5, 8), $m = -4$ **15.** (0, −4), $m = 9$

16. (1, 1), $m = 0$ **17.** (−3, −4), $m = -2$ **18.** (6, −10), $m = 5$

Write an equation in point-slope form of the line that passes through the given points.

19. (0, 0), (1, 5) **20.** (2, 3), (4, 7) **21.** (9, 6), (5, −6)

22. (8, −7), (9, −8) **23.** (1, −2), (2, 5) **24.** (2, −7), (−4, −10)

Classified Ads **In Exercises 25 and 26, use the following information.**

It costs $2.00 per day to place a one-line ad in the classifieds plus a flat service fee. One day costs $3.00 and four days costs $9.00.

25. Write a linear equation that gives the cost in dollars, y, in terms of the number of days the ad appears, x.

26. Find the cost of a six-day ad.

Travel **In Exercises 27 and 28, use the following information.**

You are driving from Grand Rapids, Michigan, to Detroit, Michigan. You leave Grand Rapids at 4:00 P.M. At 5:10 P.M. you pass through Lansing, Michigan, a distance of 65 miles.

27. Write a linear equation that gives the distance in miles, d, in terms of time, t. Let t represent the number of minutes since 4:00 P.M.

28. Approximately what time will you arrive in Detroit if it is 150 miles from Grand Rapids?

LESSON **5.5**

NAME ______________________________ DATE __________

Practice B

For use with pages 300–306

Write an equation in point-slope form of the line.

1.

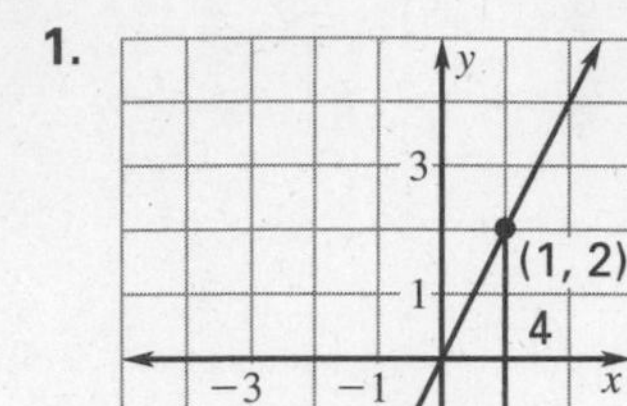

2.

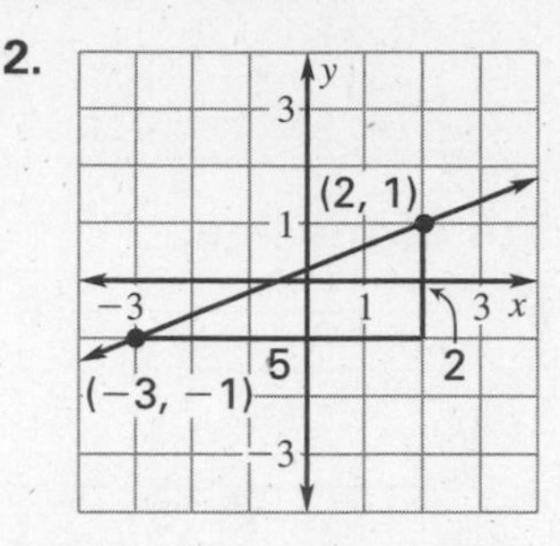

3.

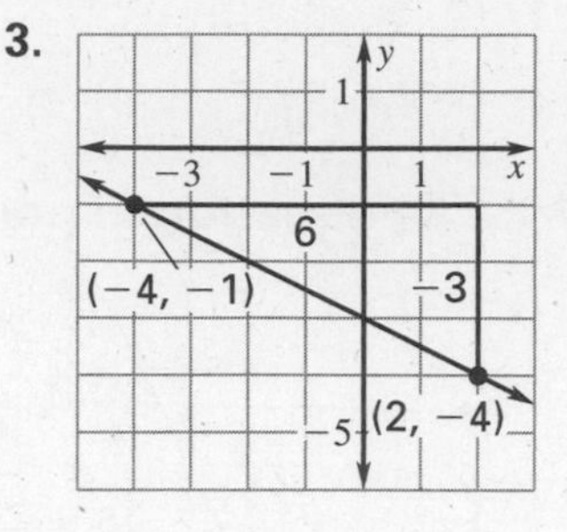

Write an equation in point-slope form of the line that passes through the given point and has the given slope.

4. $(-3, 24), m = -2$ **5.** $(-4, -2), m = -5$ **6.** $(0, -3), m = \frac{2}{3}$

7. $(6, -5), m = -4$ **8.** $(-7, 6), m = 0$ **9.** $(-3, -5), m = 6$

Write an equation in point-slope form of the line that passes through the given points.

10. $(0, 5), (1, 3)$ **11.** $(2, 4), (6, 2)$ **12.** $(3, 0), (0, -3)$

13. $(6, -2), (10, 1)$ **14.** $(-2, -3), (4, 1)$ **15.** $(1, 3), (-5, -3)$

16. $(-5, -7), (-3, -10)$ **17.** $(6, 11), (-1, 2)$ **18.** $(-3, -8), (2, 4)$

Rewrite the equation in slope-intercept form.

19. $y + 4 = 5(x + 2)$ **20.** $y - 3 = -2(x + 1)$ **21.** $y - 5 = 3(x - 4)$

22. $y + 11 = -3(x - 9)$ **23.** $y + 6 = \frac{1}{2}(x - 12)$ **24.** $y - \frac{2}{3} = 4\left(x + \frac{5}{12}\right)$

Classified Ads **In Exercises 25 and 26, use the following information.**

It costs $1.50 per day to place a one-line ad in the classifieds plus a flat service fee. One day costs $3.50 and four days costs $8.00.

25. Write a linear equation that gives the cost in dollars, y, in terms of the number of days the ad appears, x.

26. Find the cost of a six-day ad.

Travel **In Exercises 27 and 28, use the following information.**

You are flying from Houston to Chicago. You leave Houston at 7:30 A.M. At 8:35 A.M. you fly over Little Rock, a distance of 455 miles.

27. Write a linear equation that gives the distance in miles, y, in terms of time, x. Let x represent the number of minutes since 7:30 A.M.

28. Approximately what time will you arrive in Chicago if it is 950 miles from Houston?

LESSON 5.5

NAME ______________________ DATE ____________

Practice C

For use with pages 300–306

Write an equation in point-slope form of the line.

1.

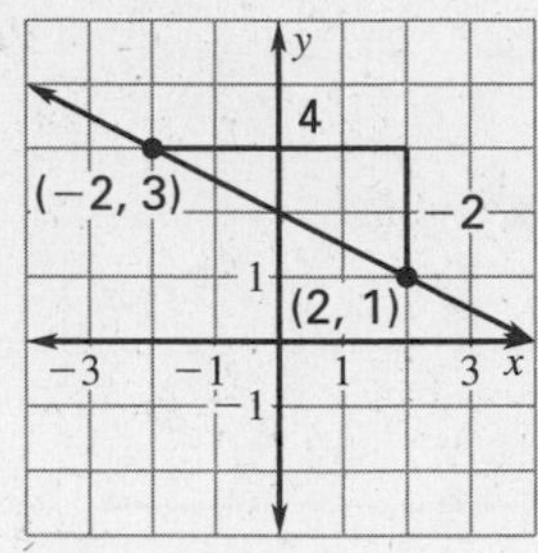

2.

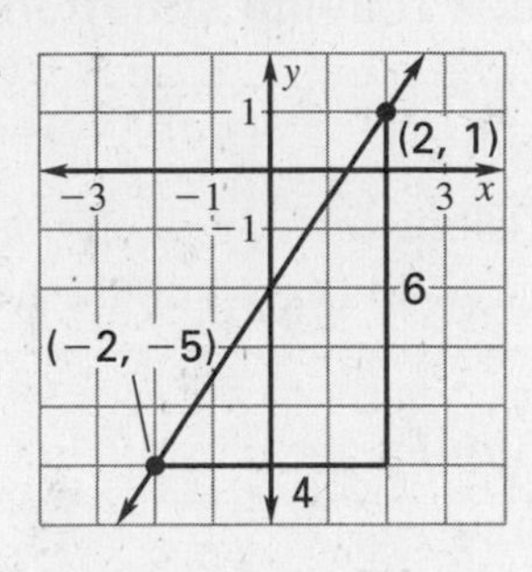

3.

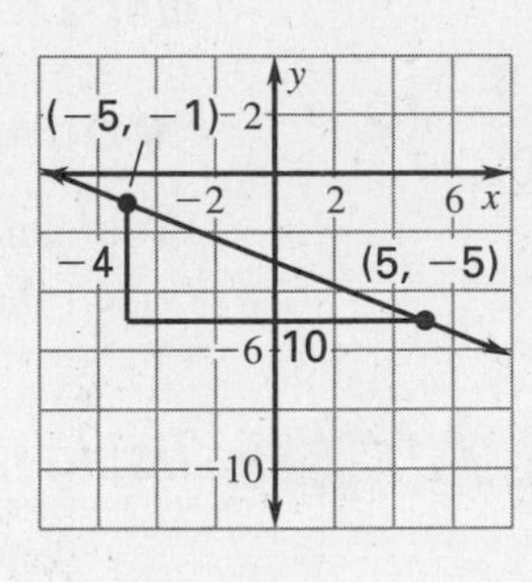

Write an equation in point-slope form of the line that passes through the given point and has the given slope.

4. $(-4, 16), m = 8$

5. $(21, -6), m = -\frac{3}{2}$

6. $(-8, -9), m = 0$

7. $(18, -35), m = \frac{7}{9}$

8. $(-12, -13), m = -\frac{1}{4}$

9. $(-10, 3), m = \frac{6}{5}$

Write an equation in point-slope form of the line that passes through the given points.

10. $(-12, 6), (4, -10)$

11. $(8, -6), (5, 21)$

12. $(24, 7), (12, 15)$

13. $(-20, 9), (-6, -9)$

14. $(-18, 8), (-11, 19)$

15. $(-40, 15), (-2, 27)$

Rewrite the equation in slope-intercept form.

16. $y - 9 = 8(x + 1)$

17. $y + 6 = -(x - 17)$

18. $y - 4 = -\frac{1}{2}(x + 14)$

19. $y - \frac{5}{6} = 7\left(x - \frac{2}{3}\right)$

20. $y + 3.4 = 2(x - 4.2)$

21. $y + 2.5 = -6(x + 1.2)$

Classified Ads **In Exercises 22 and 23, use the following information.**

It costs $3.25 per day to place a two-line ad in the classifieds plus a flat service fee. One day costs $6.75 and four days costs $16.50.

22. Write a linear equation that gives the cost in dollars, y, in terms of the number of days the ad appears, x.

23. Find the cost of a seven-day ad.

Travel **In Exercises 24 and 25, use the following information.**

You are driving from Austin, Texas, to Dallas, Texas. You leave Austin at 1:30 P.M. At 3:05 P.M. you pass through Waco, Texas, a distance of 100 miles.

24. Write a linear equation that gives the distance in miles, d, in terms of time, t. Let t represent the number of minutes since 1:30 P.M.

25. Approximately what time will you arrive in Dallas if it is 190 miles from Austin?

26. ***Wrestling Team*** The wrestling team runs 10 laps at the beginning of each practice. For every minute a team member is late for practice, he must run additional laps. Chen was 5 minutes late for practice and had to run a total of 20 laps. Write a linear equation that gives the number of laps run, y, in terms of the number of minutes late for practice, x.

NAME ______________________________ DATE ____________

Reteaching with Practice

For use with pages 300–306

GOAL **Use the point-slope form to write an equation of a line and use the point-slope form to model a real-life situation**

> **VOCABULARY**
>
> You can use the **point-slope form,** $y - y_1 = m(x - x_1)$, when you are given the slope m and a point (x_1, y_1) on the line.

EXAMPLE 1 *Using the Point-Slope Form*

Use the point-slope form of a line to write an equation of the line that passes through the points $(3, -1)$ and $(-3, 5)$.

SOLUTION

Use the points $(x_1, y_1) = (3, -1)$ and $(x_2, y_2) = (-3, 5)$ to find the slope.

$$m = \frac{y_2 - y_1}{x_2 - x_1}$$

$$= \frac{5 - (-1)}{-3 - 3}$$

$$= \frac{6}{-6} = -1$$

Use the slope and the point $(3, -1)$ as (x_1, y_1) in the point-slope form.

$y - y_1 = m(x - x_1)$	Write point-slope form.
$y - (-1) = -1(x - 3)$	Substitute for m, x_1, and y_1.
$y + 1 = -1(x - 3)$	Simplify.
$y + 1 = -x + 3$	Use distributive property.
$y = -x + 2$	Subtract 1 from each side.

Note: You can use either point as (x_1, y_1).

Exercises for Example 1

Use the point-slope form of a line to write an equation of the line that passes through the given points.

1. $(4, 5), (6, 9)$ **2.** $(-1, 6), (0, 3)$ **3.** $(-2, 8), (2, -8)$

NAME ______________________ DATE ____________

Reteaching with Practice

For use with pages 300–306

EXAMPLE 2 Writing and Using a Linear Model

You are running a 10-kilometer race. At 8:00 A.M., you start the race. At 8:30 A.M., you are 4 kilometers from the finish line. Write a linear model that gives the distance d (in kilometers) from the starting line in terms of the time t (in minutes). Let t represent the number of minutes since 8:00 A.M.

SOLUTION

One point on the line is $(t_1, d_1) = (0, 10)$. Another point on the line is $(t_2, d_2) = (30, 4)$. Find the slope of the line.

$$m = \frac{\text{change in distance}}{\text{change in time}} = \frac{4 - 10}{30 - 0} = \frac{-6}{30} = -0.2$$

Use the point-slope form to write the model. Use $(0, 10)$ as (t_1, d_1).

$d - d_1 = m(t - t_1)$ Write point-slope form.

$d - 10 = -0.2(t - 0)$ Substitute for m, d_1, and t_1.

$d - 10 = -0.2t + 0$ Use distributive property.

$d = -0.2t + 10$ Add 10 to each side.

Exercises for Example 2

4. Use the linear model from Example 2 to predict your time to finish the race.

5. Rework Example 2 if at 8:30 A.M., you are 5 kilometers from the finish line.

NAME ______________________________ DATE __________

Quick Catch-Up for Absent Students

For use with pages 300–306

The items checked below were covered in class on (date missed) __________

Lesson 5.5: Point-Slope Form of a Linear Equation

____ **Goal 1:** Use the point-slope form to write an equation of a line. (pp. 300–301)

Material Covered:

____ Example 1: Developing the Point-Slope Form

____ Study Help: Study Tip

____ Example 2: Using the Point-Slope Form

____ Activity: Investigating the Point-Slope Form

Vocabulary:

point-slope form, p. 300

____ **Goal 2:** Use the point-slope form to model a real-life situation. (p. 302)

Material Covered:

____ Example 3: Writing and Using a Linear Model

____ Other (specify) ______________________________

____

Homework and Additional Learning Support

____ Textbook (specify) pp. 303–306

____

____ Internet: Extra Examples at www.mcdougallittell.com

____ *Reteaching with Practice* worksheet (specify exercises) ____________

____ *Personal Student Tutor* for Lesson 5.5

NAME ______________________________ DATE __________

Interdisciplinary Application

For use with pages 300–306

Advertising

BUSINESS Marketing is a major factor in the success or failure of a company's product. While many people associate marketing with advertising, most fail to realize it actually consists of four key areas: the product, its price, the promotional campaign, and the distribution area. Advertising is only one facet of promotion. When advertising a product, a company will usually isolate a target audience or customer. The target is a specific group of people that would like to buy and can afford the product.

There are many products and services that are specifically aimed at the youth of America. Young adults have a lot of discretionary income. Discretionary income is money leftover after paying all necessary expenses. A problem arises when trying to market in a different culture or country.

For instance, Mexico has a large percentage of its population less than 18 years of age, and this age group is expanding. Companies that are successfully selling products to teenagers in the U.S. would love to expand their product lines into Mexico. However, they must be careful to alter their advertising campaigns to relate to the Mexican teenager who has a different background and culture than the U.S. teenager.

In Exercises 1–3, use the following information.

A Company selling blue jeans in Mexico had sales of 45 million pesos when the population of teenagers was 3 million people. As the teenage population grew to 4.5 million people, sales increased to 60 million pesos.

1. Write an equation of the line in point-slope form that passes through the above points, letting x represent population and y represent sales in millions of pesos. Use the point $(3,45)$.
2. Graph the line.
3. Predict sales when the teenage population reaches 7 million.

In Exercises 4–6, use the following information.

A music company sells compact disks and cassettes and has sales of 15 million pesos when the teenage population is 3 million. When the population grows to 4.5 million, sales increase to 26 million pesos.

4. Write an equation of the line in point-slope form that passes through the given points, letting x represent population and y represent sales in millions of pesos. Use the point $(3, 15)$.
5. Graph the line found in Exercise 4.
6. Predict sales when the teenage population reaches 6 million.

LESSON 5.5

NAME ______________________ DATE __________

Math and History Application

For use with page 306

HISTORY Willard Libby and his team of scientists at the University of Chicago first developed radiocarbon dating in 1947. In 1960 Libby received the Nobel Prize in Chemistry for this discovery.

Radiocarbon dating is based on the fact that all plants and animals are made of carbon. Libby discovered how long radiocarbon takes to decay and found the half-life of radiocarbon is about 5730 years. Half-life is the amount of time it takes for half the radiocarbon to disappear. This means that in about 11,000 years another half of the remaining amount of radiocarbon will disappear. Radiocarbon dating doesn't work with objects older than about 50,000 years because most of the radiocarbon would be gone in an object that old. In measuring the amount of radiocarbon present, scientists are able to determine the age of an object.

Prior to radiocarbon dating, archaeologists often used relative dating. Relative dating is based on the age of other materials found at a site. Archaeologists thought that the same kinds of artifacts would be the same age. Where an object was found was also important; an artifact found in a layer below another was considered older. Relative dating was not very accurate and scientists were unable to date all artifacts.

MATH Once radiocarbon dating was discovered, it had to be tested for accuracy. One way scientists tested radiocarbon dates was to use dendrochronology, or tree ring dating. This dating method uses the annual rings a tree produces each year. By counting and measuring the rings, scientists can date the age of a tree. Scientists can then date the rings using radiocarbon testing and compare the two results. Radiocarbon dating is an accurate way to date objects, however the date may differ from the true age by a small amount.

1. In 1991 a frozen, mummified body was found in the Italian Alps. At first authorities thought the body was of a modern person. Upon closer inspection, the body appeared much older because of the types of tools near the body. Archaeologists first determined the body, which they called the Iceman, to be about 4000 years old because of the stone knife he carried. This guess proved incorrect. Radiocarbon dating determined that the Iceman lived around 3000 to 3500 B.C. According to these results, how many years ago did the Iceman live? What percent of radiocarbon remained in his body?

2. A scientist discovers that in A.D. 1000 a certain tree is 1000 radiocarbon years before present. Using A.D. 2000 for the year and 0 for the radiocarbon years before present for the second point, write the point-slope form of the equation of the line that passes through the two points. Then rewrite the equation in slope-intercept form. Using tree ring dating, the scientist finds the age of the tree to be 500 radiocarbon years before present in the year A.D. 1500. Does this point fall on the line you found for the radiocarbon dating?

NAME ______________________________ DATE __________

Challenge: Skills and Applications

For use with pages 300–306

In Exercises 1–8, write an equation in point-slope form of the line that passes through the given points.

1. $\left(\frac{1}{2}, -4\right), (2, 11)$
2. $(8, 3), \left(-\frac{1}{3}, -2\right)$
3. $(-0.5, 0.9), (-3.3, -0.5)$
4. $(3.2, -1.4), (2.4, 1.8)$
5. $\left(\frac{3}{2}, -\frac{1}{3}\right), \left(-\frac{2}{3}, 4\right)$
6. $(5, -4), \left(\frac{1}{2}, -\frac{1}{4}\right)$
7. $(p, q), (-p, 2q)$
8. $(2p, -q), (p, p - q)$

In Exercises 9–12, use the following information.

A line passes through the point $(6, 3)$ and has slope $-\frac{5}{2}$.

9. Write an equation of the line in point-slope form.
10. For the given point $(6, 3)$, the x-coordinate is twice the y-coordinate. Find a point on the line for which the y-coordinate is twice the x-coordinate. Explain your method.
11. Find a point on the line for which the two coordinates are opposites.
12. Find a point on the line for which the sum of the two coordinates is 15.

In Exercises 13–15, use the following information.

Hiking up a mountain, Zahara looked at a topographical map and saw that at 1:00 P.M. her elevation was 5620 feet above sea level. By 2:30 P.M. she had reached an elevation of 6040 feet above sea level.

13. Write an equation in point-slope form that gives Zahara's elevation y, at a time x hours after noon.
14. If Zahara continues at the same rate, what elevation can she expect to reach by 5:00 P.M.?
15. Use the model from Exercise 13 to find Zahara's elevation at 10:00 A.M.

NAME ______________________________ DATE __________

Quiz 2

For use after Lessons 5.4–5.5

Answers

1. ____________

2. ____________

3. ____________

4. ____________

5. ____________

1. Draw a best-fitting line for the scatter plot. Write an equation of your line. *(Lesson 5.4)*

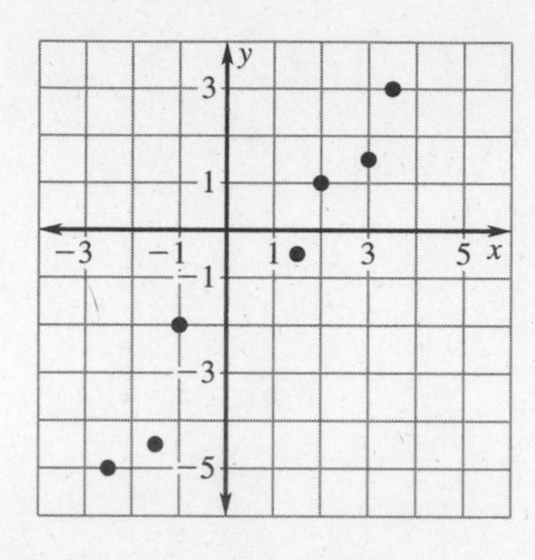

2. Draw a scatter plot of the data. If possible, draw a best-fitting line for the scatter plot and write an equation of the line. State whether x and y have a *positive correlation*, a *negative correlation*, or *no correlation*. *(Lesson 5.4)*

x	y
-2.5	4.4
3.5	-0.4
0.5	1
1.6	1.7
-2	3
-1.3	4

3. Write an equation of the line in point-slope form. *(Lesson 5.5)*

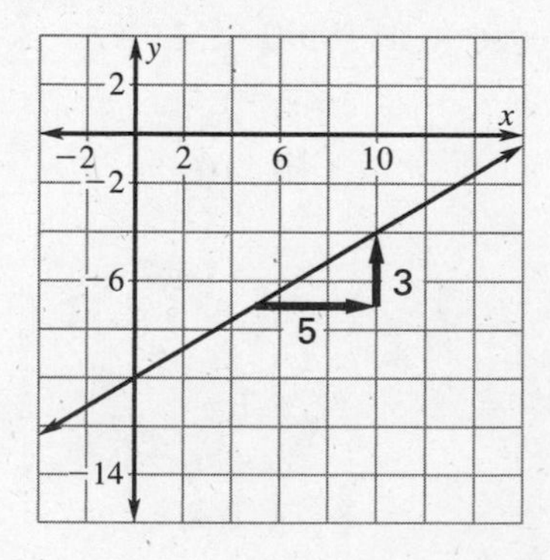

4. Write an equation in point-slope form of the line that passes through $(3, 7)$ and $(-5, 4)$. *(Lesson 5.5)*

5. Write an equation in point-slope form of the line that passes through $(-7, 1)$ and has a slope of $m = \frac{1}{2}$. *(Lesson 5.5)*

LESSON 5.6

TEACHER'S NAME ______________________ CLASS ______ ROOM ______ DATE ______

Lesson Plan

2-day lesson (See *Pacing the Chapter,* TE pages 270C–270D) For use with pages 307–314

GOALS
1. Write a linear equation in standard form.
2. Use the standard form of an equation to model real-life situations.

State/Local Objectives ____

____

Lesson 5.6

✓ Check the items you wish to use for this lesson.

STARTING OPTIONS

____ Homework Check: TE page 303; Answer Transparencies
____ Warm-Up or Daily Homework Quiz: TE pages 308 and 305, CRB page 79, or Transparencies

TEACHING OPTIONS

____ Motivating the Lesson: TE page 309
____ Concept Activity: SE page 307; CRB page 80 (Activity Support Master)
____ Lesson Opener (Activity): CRB page 81 or Transparencies
____ Graphing Calculator Activity with Keystrokes: CRB pages 82–83
____ Examples: Day 1: 1–3, SE pages 308–309; Day 2: 4, SE page 310
____ Extra Examples: Day 1: TE page 309 or Transp.; Day 2: TE page 310 or Transp.
____ Closure Question: TE page 310
____ Guided Practice: SE page 311; Day 1: Exs. 1–15; Day 2: Exs. 16–17

APPLY/HOMEWORK

Homework Assignment

____ Basic Day 1: 18–52 even; Day 2: 54–69, 75–80, 86, 91, 96, 101
____ Average Day 1: 18–52 even; Day 2: 54–69, 75–80, 86, 91, 96, 104
____ Advanced Day 1: 18–52 even; Day 2: 54–69, 75–83, 86, 91, 96, 101

Reteaching the Lesson

____ Practice Masters: CRB pages 84–86 (Level A, Level B, Level C)
____ Reteaching with Practice: CRB pages 87–88 or Practice Workbook with Examples
____ Personal Student Tutor

Extending the Lesson

____ Applications (Real-Life): CRB page 90
____ Challenge: SE page 314; CRB page 91 or Internet

ASSESSMENT OPTIONS

____ Checkpoint Exercises: Day 1: TE page 309 or Transp.; Day 2: TE page 310 or Transp.
____ Daily Homework Quiz (5.6): TE page 314, CRB page 94, or Transparencies
____ Standardized Test Practice: SE page 313; TE page 314; STP Workbook; Transparencies

Notes ____

____

____

TEACHER'S NAME ______________________ CLASS ______ ROOM ______ DATE ______

Lesson Plan for Block Scheduling

1-day lesson (See *Pacing the Chapter,* TE pages 270C–270D) For use with pages 307–314

GOALS
1. **Write a linear equation in standard form.**
2. **Use the standard form of an equation to model real-life situations.**

State/Local Objectives ______________________

____

____

CHAPTER PACING GUIDE	
Day	**Lesson**
1	5.1 (all); 5.2 (all)
2	5.3 (all)
3	5.4 (all); 5.5 (begin)
4	5.5 (end); **5.6 (begin)**
5	**5.6 (end)**; 5.7 (begin)
6	5.7 (end); Review Ch. 5
7	Assess Ch. 5; 6.1 (all)

✓ Check the items you wish to use for this lesson.

STARTING OPTIONS

____ Homework Check: TE page 303; Answer Transparencies
____ Warm-Up or Daily Homework Quiz: TE pages 308 and 305, CRB page 79, or Transparencies

TEACHING OPTIONS

____ Motivating the Lesson: TE page 309
____ Concept Activity: SE page 307; CRB page 80 (Activity Support Master)
____ Lesson Opener (Activity): CRB page 81 or Transparencies
____ Graphing Calculator Activity with Keystrokes: CRB page 82–83
____ Examples: Day 4: 1–3, SE pages 308–309; Day 5: 4, SE page 310
____ Extra Examples: Day 4: TE page 309 or Transp.; Day 5: TE page 310 or Transp.
____ Closure Question: TE page 310
____ Guided Practice: SE page 311; Day 4: Exs. 1–15; Day 5: Exs. 16–17

APPLY/HOMEWORK

Homework Assignment (See also the assignments for Lessons 5.5 and 5.7.)
____ Block Schedule: Day 4: 18–52 even; Day 5: 54–69, 75–84, 90, 95, 100, 105

Reteaching the Lesson
____ Practice Masters: CRB pages 84–86 (Level A, Level B, Level C)
____ Reteaching with Practice: CRB pages 87–88 or Practice Workbook with Examples
____ Personal Student Tutor

Extending the Lesson
____ Applications (Real-Life): CRB page 90
____ Challenge: SE page 314; CRB page 91 or Internet

ASSESSMENT OPTIONS

____ Checkpoint Exercises: Day 4: TE page 309 or Transp.; Day 5: TE page 310 or Transp.
____ Daily Homework Quiz (5.6): TE page 314, CRB page 94, or Transparencies
____ Standardized Test Practice: SE page 313; TE page 314; STP Workbook; Transparencies

Notes ______________________

____

____

NAME ______________________ DATE __________

Available as a transparency

WARM-UP EXERCISES

For use before Lesson 5.6, pages 307–314

Write an equation for each line in point-slope form.

1. the line through $(-1, 3)$ with slope -2
2. the line through $(3, -5)$ and $(2, -4)$

Convert each equation in point-slope form to slope-intercept form.

3. $y + 1 = -2(x - 5)$
4. $y - 3 = \frac{1}{2}(x + 6)$

DAILY HOMEWORK QUIZ

For use after Lesson 5.5, pages 300–306

Write an equation in point-slope form of the line that passes through the points.

1. $(5, 3), (-1, 0)$
2. $(-4, -2), (1, -17)$
3. What is an equation in point-slope form of the line through $(6, 4)$ that is perpendicular to the line $y - 2 = \frac{2}{5}(x - 5)$?

NAME ______________________ DATE __________

Activity Support Master

For use with page 307

Step 1

Number of GFE shares	0	1	2	3	4	5
Number of JIH shares	1	1	1	1	1	1
Total cost (dollars)						

Step 2

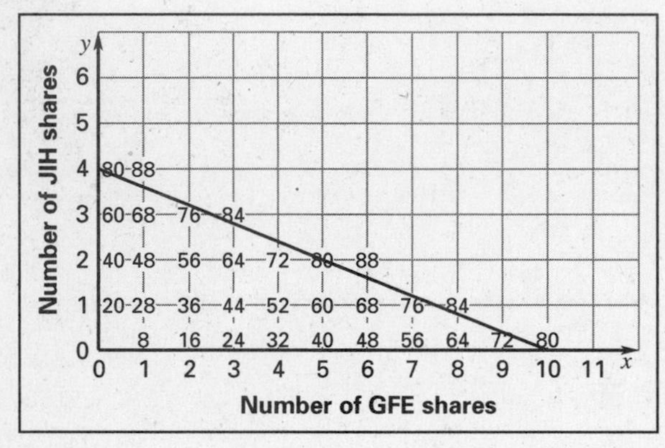

Step 3

____

____

____

____

Step 4

____

NAME ______________________ DATE ____________

Available as a transparency

Activity Lesson Opener

For use with pages 308–314

SET UP: Work with a group.

YOU WILL NEED: • 16 index cards

1. Write each equation on a separate index card.

$y = 2x + 1$	$-x + 5y = 10$
$y = 4x - 1$	$x - y = 6$
$y = \frac{1}{5}x + 2$	$x + 2y = -8$
$y = -x + 3$	$-2x + y = 1$
$y = -\frac{1}{2}x - 4$	$x + y = 3$
$y = 5x + 1$	$4x - y = 1$
$y = x - 6$	$5x - y = -1$
$y = \frac{2}{3}x + 3$	$-2x + 3y = 9$

2. Use your cards to play "Equation Concentration." Shuffle the cards and place them face down on a desk or table. Choose a member of your group to go first. The first player turns over two cards. If the player thinks that the equations on the cards represent equivalent equations, he or she picks up the two cards. If not, he or she turns the cards over and returns them to the bottom of the deck. The player to the right goes next. If two cards are picked up and any other player thinks the equations on the two cards are not equivalent, he or she can call for a "challenge." To win a "challenge," a player must prove that the equations are not equivalent. The player who loses a "challenge" loses his or her next turn. A player's turn ends when he or she loses a challenge or when two cards are turned over that do not match. The game is over when all the cards have been matched. The player with the most cards at the end of the game "wins."

NAME ______________________ DATE __________

Graphing Calculator Activity

For use with pages 308–314

GOAL **To determine if two linear equations are equivalent**

Two equations are equivalent if they have the same solution set. A table of solutions can take some time to construct. A graphing calculator, however, allows you to easily view a table of solutions for an equation.

X	Y1	
0	1	
1	3	
2	5	
3	7	
4	9	
5	11	
6	13	

Activity

1. Copy and complete the table of solutions for the equation $2x + y = 1$.

x	-2	-1	0	1	2
y					

2. Enter the equation $y = -2x + 1$ into your graphing calculator.
3. Compare the table of solutions for $y = -2x + 1$ generated by your calculator with the table you completed in Step 1.
4. Can you conclude that the two equations are equivalent?
5. Repeat Steps 1–4 for the equations $3x - 4y = 6$ and $y = \frac{3}{4}x + 6$.

Exercises

1. For the first equation, make a table of solutions. Enter the second equation into your graphing calculator and view the table generated. Compare the two tables. Can you conclude that the two equations are equivalent?

a. $x + y = 3$
$y = -x + 3$

b. $5x - 6y = 10$
$y = \frac{5}{6}x + 10$

c. $7x - 8y = -2$
$y = \frac{7}{8}x + \frac{1}{4}$

2. Are you able to enter the equation $4x + y = 9$, as it is written, into your graphing calculator and plot the graph? If not, what could you do to get your calculator to graph the equation?

NAME ______________________________ DATE ____________

Graphing Calculator Activity

For use with pages 308–314

TI-82

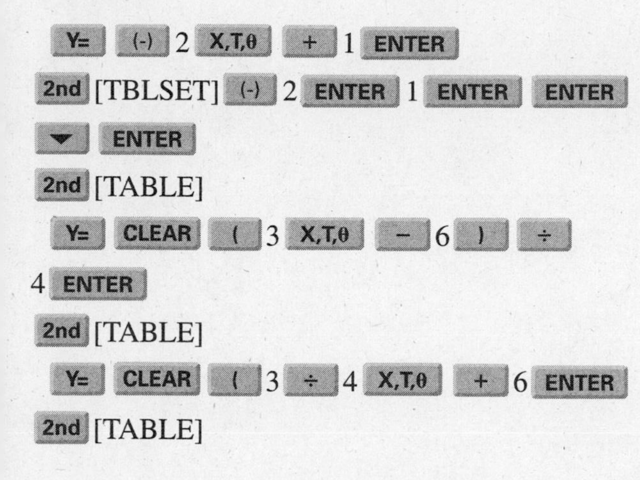

TI-83

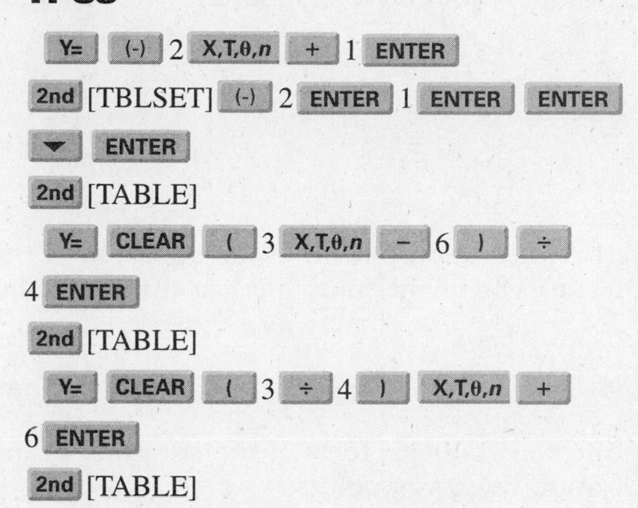

SHARP EL-9600c

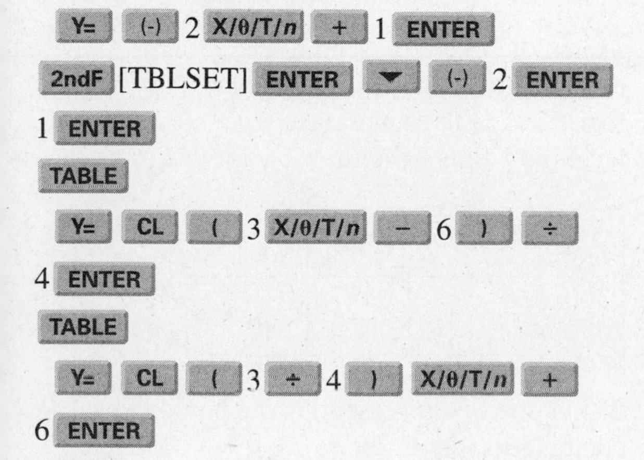

CASIO CFX-9850GA PLUS

From the main menu, choose TABLE.

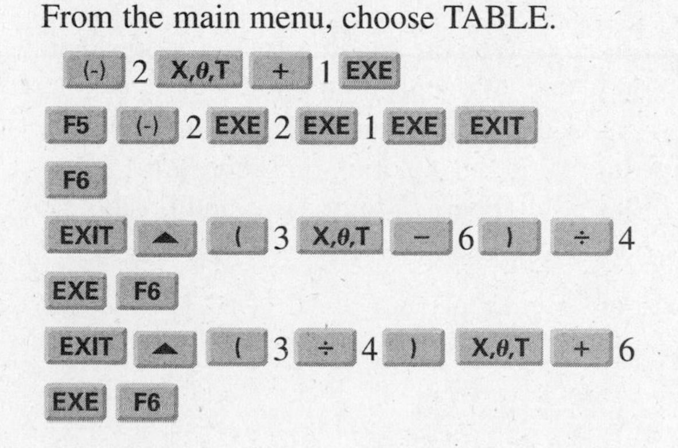

LESSON 5.6

NAME ______________________________ DATE ____________

Practice A

For use with pages 308–314

Write the equation in standard form with integer coefficients.

1. $x - y - 9 = 0$

2. $-4y + 6x + 7 = 0$

3. $x + 7 = 0$

4. $3y + 2x = 6$

5. $y = -11x - 4$

6. $y - 1 = 0$

7. $3 + 4x - y = 0$

8. $x - 8y + 2 = 0$

9. $x = y$

10. $y = 5x - \frac{1}{2}$

11. $y = \frac{1}{4}x + 3$

12. $y = -\frac{2}{3}x - 1$

Write the standard form of the equation of the line that passes through the given point and has the given slope.

13. $(0, 4), m = 1$

14. $(2, 5), m = -3$

15. $(-1, 3), m = 8$

16. $(6, -7), m = 4$

17. $(5, 6), m = -2$

18. $(-4, -9), m = -2$

Write the standard form of the equation of the line that passes through the given points.

19. $(2, -5), (8, 1)$

20. $(-1, -2), (0, 3)$

21. $(1, -6), (-5, 6)$

22. $(3, 19), (-2, -11)$

23. $(-4, 3), (-1, -6)$

24. $(2, 18), (-2, 2)$

25. ***Publicity*** You are running for class president. You have \$30 to spend on publicity. It costs \$2 to make a campaign button and \$1 to make a poster. Write an equation that represents the different numbers of buttons, x, and posters, y, you could make.

26. Sketch the line representing the possible combinations of buttons and posters in Exercise 25. Then complete the table and label the points from the table on the graph.

Number of buttons	0	5	8	10	15
Number of posters					

27. ***Canning Jelly*** Your grandmother made 240 oz. of jelly. You have two types of jars. The first holds 10 oz. and the second holds 12 oz. Write an equation that represents the different numbers of 10-oz. jars, x, and 12-oz. jars, y, that will hold all of the jelly.

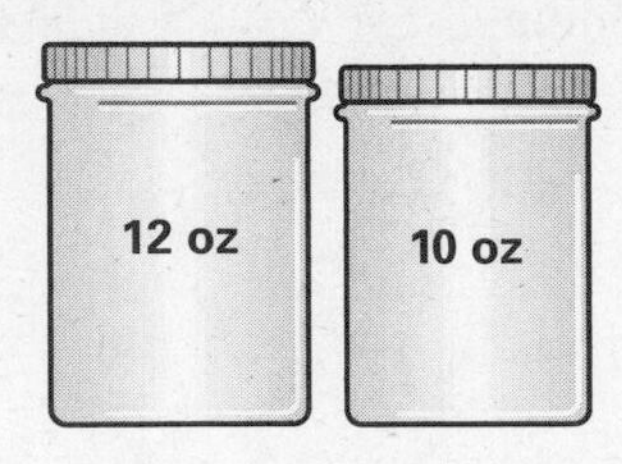

28. Sketch the line representing the possible jar combinations in Exercise 27. Then complete the table and label the points from the table on the graph.

10-oz. jars	0	6	12	18	24
12-oz. jars					

NAME ______________________ DATE __________

Practice B

For use with pages 308–314

Write the equation in standard form with integer coefficients.

1. $2x - y - 8 = 0$
2. $0.3x - 0.4y = 7.5$
3. $y = 3x + 2$
4. $y = 5 - 3x$
5. $0.6x = 2.1y + 1.8$
6. $2x = 3y + 5$
7. $x - 4 = 0$
8. $3y = 12$
9. $2x - 9 = \frac{3}{5}y$
10. $\frac{1}{4}x - 2y = -3$
11. $y = \frac{1}{2}x + 4$
12. $y = \frac{2}{3}x - \frac{5}{3}$

Write the standard form of the equation of the line that passes through the given point and has the given slope.

13. $(4, 3), m = 2$
14. $(1, 5), m = -4$
15. $(0, 6), m = 3$
16. $(-2, 4), m = -6$
17. $(6, -8), m = \frac{1}{3}$
18. $(-2, 4), m = -\frac{1}{2}$

Write the standard form of the equation of the line that passes through the given points.

19. $(5, 8), (3, 2)$
20. $(-2, 5), (3, -10)$
21. $(-7, 3), (1, 2)$
22. $(-4, -5), (-2, 5)$
23. $(8, 1), (4, -1)$
24. $(-6, 6), (3, 3)$

Write the standard form of the equation of the horizontal and vertical lines that pass through the given point.

25. $(3, -4)$
26. $(5, 1)$
27. $(-3, -2)$
28. $(0, -4)$

Party Food **In Exercises 29–32, use the following information.**

You are in charge of buying the hamburger and boned chicken for a party. You have \$60 to spend. The hamburger costs \$2 per pound and boned chicken is \$3 per pound.

29. Write an equation that represents the different amounts of hamburger, x, and chicken, y, that you can buy.
30. Rewrite the equation in Exercise 29 in slope-intercept form.
31. Sketch the graph of the linear equation in Exercise 29.
32. Complete the table and label the points from the table on the graph.

Hamburger (lb), x	0	6	12	18	30
Chicken (lb), y					

Lawn Seed **In Exercises 33–36, use the following information.**

You are buying \$48 worth of lawn seed that consists of two types of seed. One type is a quick-growing rye grass that costs \$4 per pound, and the other type is a higher-quality seed that costs \$6 per pound.

33. Write an equation that represents the different amounts of \$4 seed, x, and \$6 seed, y, that you can buy.
34. Rewrite the equation in Exercise 33 in slope-intercept form.
35. Sketch the graph of the linear equation in Exercise 33.
36. Complete the table and label the points from the table on the graph.

\$4 seed (lb), x	0	3	6	9	12
\$6 seed (lb), y					

LESSON **5.6**

NAME ______________________ DATE __________

Practice C

For use with pages 308–314

Write the equation in standard form with integer coefficients.

1. $-3x + y + 9 = 0$
2. $y = 6x - 12$
3. $5y - 4x + 40 = 0$
4. $0.8x = -9.5 - 0.7y$
5. $-9.6y - 6.7x + 4.2 = 0$
6. $-3.5y = 7x - 6.8$
7. $\frac{1}{4}x + 16 = 0$
8. $-\frac{15}{7} = 5y$
9. $\frac{5}{9}y + \frac{2}{9}x - \frac{8}{9} = 0$
10. $-\frac{4}{5}x = -5y + \frac{1}{3}$
11. $y = \frac{3}{4}x - 20$
12. $\frac{2}{3}x + \frac{7}{9}y + 12 = 0$

Write the standard form of the equation of the line that passes through the given point and has the given slope.

13. $(6, 4), m = 5$
14. $(-3, -8), m = -3$
15. $(-2, 15), m = -6$
16. $(-2, 4), m = -\frac{1}{6}$
17. $(5, -4), m = \frac{1}{3}$
18. $(-3, 4), m = -\frac{1}{2}$

Write the standard form of the equation of the line that passes through the given points.

19. $(3, -7), (-3, -11)$
20. $(-7, 7), (14, -5)$
21. $(8, 2), (17, 9)$
22. $(1, 0), (13, 15)$
23. $(0, -3), (7, -5)$
24. $(2, -3), (-6, 2)$

Write the standard form of the equation of the horizontal and vertical lines that pass through the given point.

25. $\left(0, \frac{1}{2}\right)$
26. $(-7, 4)$
27. $(-8, -10)$
28. $(5, -3)$

Party Food **In Exercises 29–33, use the following information.**

You are in charge of buying the hamburger and boned chicken for a party. You have $60 to spend. The hamburger costs $1.50 per pound and boned chicken is $2.50 per pound.

29. Write an equation that represents the different amounts of hamburger, *x*, and chicken, *y*, that you can buy.
30. Rewrite the equation in Exercise 29 in slope-intercept form.
31. Sketch the graph of the linear equation in Exercise 29.
32. Complete the table and label the points from the table on the graph.

Hamburger (lb), x	0	5	10	25	40
Chicken (lb), y					

33. Name the *x*- and *y*-intercepts. Interpret their meaning in the context of the problem.

Lawn Seed **In Exercises 34–38, use the following information.**

You are buying $75 worth of lawn seed that consists of two types of seed. One type is a quick-growing rye grass that costs $3.75 per pound, and the other type is a higher-quality seed that costs $5 per pound.

34. Write an equation that represents the different amounts of $3.75 seed, *x*, and $5 seed, *y*, that you can buy.
35. Rewrite the equation in Exercise 34 in slope-intercept form.
36. Sketch the graph of the linear equation in Exercise 34.
37. Complete the table and label the points from the table on the graph.

$3.75 seed (lb), x	0	4	8	16	20
$5 seed (lb), y					

38. Name the *x*- and *y*-intercepts. Interpret their meaning in the context of the problem.

NAME ______________________________ DATE __________

Reteaching with Practice

For use with pages 308–314

GOAL **Write a linear equation in standard form and use the standard form of an equation to model real-life situations**

VOCABULARY

The **standard form** of the equation of a line is $Ax + By = C$, where A, B, and C represent real numbers and A and B are not both zero.

EXAMPLE 1 *Writing an Equation in Standard Form*

Write $y = -\frac{3}{4}x + 5$ in standard form with integer coefficients.

SOLUTION

To write the equation in standard form, isolate the variable terms on the left and the constant term on the right.

$y = -\frac{3}{4}x + 5$	Write original equation.
$4y = 4\left(-\frac{3}{4}x + 5\right)$	Multiply each side by 4.
$4y = -3x + 20$	Use distributive property.
$3x + 4y = 20$	Add $3x$ to each side.

Exercises for Example 1

Write the equation in standard form with integer coefficients.

1. $y = \frac{2}{3}x - 7$ **2.** $y = 8 + 2x$ **3.** $y = 6 - \frac{1}{4}x$

EXAMPLE 2 *Writing a Linear Equation*

Write the standard form of the equation passing through (3, 7) with a slope of 2.

SOLUTION

Write the point-slope form of the equation of the line.

$y - y_1 = m(x - x_1)$	Write point-slope form.
$y - 7 = 2(x - 3)$	Substitute for y_1, m, and x_1.
$y - 7 = 2x - 6$	Use distributive property.
$-2x + y = 1$	Add $-2x$ and 7 to each side.

NAME ______________________________ DATE ____________

Reteaching with Practice

For use with pages 308–314

Exercises for Example 2

Write the standard form of the equation of the line that passes through the given point and has the given slope.

4. $(1, 4),\ m = -2$ **5.** $(-3, 1),\ m = 3$ **6.** $(5, -2),\ m = -1$

EXAMPLE 3 Writing and Using a Linear Model

You have \$12 to buy peaches and blueberries for a fruit salad. Peaches cost \$1.50 per pound and blueberries cost \$4.00 per pound. Write a linear equation that models the different amounts of peaches x and blueberries y that you can buy.

SOLUTION

Verbal Model [Price of peaches] · [Weight of peaches] + [Price of blueberries] · [Weight of blueberries] = [Total cost]

Labels

Price of peaches = 1.50	(dollars per pound)
Weight of peaches = x	(pounds)
Price of blueberries = 4	(dollars per pound)
Weight of blueberries = y	(pounds)
Total cost = 12	(dollars)

Algebraic Model $1.50x + 4y = 12$ Linear model

Exercise for Example 3

7. Copy and complete the table using the linear model in Example 3.

Peaches (lb), x	0	2	4	8
Blueberries (lb), y	?	?	?	?

LESSON 5.6

NAME ______________________ DATE __________

Quick Catch-Up for Absent Students

For use with pages 307–314

The items checked below were covered in class on (date missed) __________

Activity 5.6: Investigating the Standard Form of a Linear Equation (p. 307)

____ **Goal:** Model the possible combinations of two stocks that can be purchased with a limited amount of money.

Lesson 5.6: The Standard Form of a Linear Equation

____ **Goal 1:** Write a linear equation in standard form. (pp. 308–309)

Material Covered:

____ Activity: Investigating Forms of Equations

____ Example 1: Writing an Equation in Standard Form

____ Student Help: Look Back

____ Example 2: Writing a Linear Equation

____ Example 3: Horizontal and Vertical Lines

Vocabulary:

standard form, p. 308

____ **Goal 2:** Use the standard form of an equation to model real-life situations. (p. 310)

Material Covered:

____ Example 4: Writing and Using a Linear Model

____ Other (specify) ______________________

____

Homework and Additional Learning Support

____ Textbook (specify) pp. 311–314

____

____ *Reteaching with Practice* worksheet (specify exercises) __________

____ *Personal Student Tutor* for Lesson 5.6

Lesson 5.6

NAME ______________________ DATE __________

Real-Life Application: When Will I Ever Use This?

For use with pages 308–314

Saving Money

The best way to get in the habit of saving money is to start young and be consistent. Even a small amount of money saved each month will add up over time. Once people start saving, most quickly realize they do not miss money.

In Exercises 1–3, use the following information.

Mitch wants to save some money to purchase his own car. He gets two part-time jobs during summer vacation. The first, working as a busboy at a local restaurant, pays $6.00 per hour. The second, a gas station attendant position, pays $5.50 per hour. He would like to earn $200 per week.

1. Write an equation in standard form that models the different amounts of time he can work each job.
2. Graph the equation.
3. If Mitch saves 60% of his income, how much will he have saved by the end of 12 weeks of summer vacation?

In Exercises 4–5, use the following information.

Halie wants a phone line in her room. Her parents say she can have her own line if she pays for the hookup, the phone, and the monthly bill. They decide she must have $300 saved to cover all these costs. She earns $3 an hour babysitting and $2 an hour for household chores.

4. Write an equation in standard form that models the number of hours babysitting b and hours of chores c completed to raise the required money.
5. Graph the equation.

LESSON 5.6

NAME ______________________ DATE __________

Challenge: Skills and Applications

For use with pages 308–314

In Exercises 1–4, write an equation in standard form of the line that passes through the two points.

Example: $\left(\frac{1}{4}, \frac{3}{8}\right), \left(\frac{3}{5}, \frac{1}{2}\right)$

Solution:

$y - \frac{1}{2} = \frac{5}{14}\left(x - \frac{3}{5}\right)$	Find slope and write equation in point-slope form.
$y - \frac{1}{2} = \frac{5}{14}x - \frac{3}{14}$	Use distributive property.
$14\left(y - \frac{1}{2}\right) = 14\left(\frac{5}{14}x - \frac{3}{14}\right)$	Multiply each side by least common denominator.
$14y - 7 = 5x - 3$	Use distributive property.
$-5x + 14y = 4$	Add $-5x$ and 7 to each side.

1. $\left(\frac{1}{2}, -\frac{2}{3}\right), \left(\frac{3}{4}, \frac{7}{8}\right)$
2. $\left(-3, 2\frac{1}{4}\right), \left(3\frac{1}{5}, 10\right)$
3. $(-1, p), (2, -4)$
4. $(3, -2), (1, q)$

In Exercises 5–7, use the following information.

Melissa Jenkins is trying to determine the composition of a 12-milligram chemical solution. She knows the solution contains two chemicals. Each milliliter of one chemical weighs 2 milligrams and each milliliter of the other weighs 3 milligrams.

5. Write an equation that represents the different numbers of milliliters of each chemical that could be in the solution.
6. If there are 1.8 milliliters of the first chemical, how much of the second chemical is there?
7. Is it possible that the mixture contains the same number of milliliters of each chemical? If so, what is that number of milliliters?

In Exercises 8–9, Steven Chang is cutting shelves that are either 4.5 feet or 3.75 feet long. (Ignore the width of the saw blade.)

8. Write an equation that represents the numbers of shelves of each length that Steven can cut if he has 24 feet of wood. How many shelves of each length can he cut with no wood left over?
9. The wood is only sold in lengths of 12 feet or less. Will Steven need to piece together sections of wood to make a shelf? Explain.

TEACHER'S NAME ______ CLASS ______ ROOM ______ DATE ______

Lesson Plan

2-day lesson (See *Pacing the Chapter,* TE pages 270C–270D) For use with pages 315–322

1. **Determine whether a linear model is appropriate.**
2. **Use a linear model to make a real-life prediction.**

State/Local Objectives ______

✓ Check the items you wish to use for this lesson.

STARTING OPTIONS

____ Homework Check: TE page 311; Answer Transparencies
____ Warm-Up or Daily Homework Quiz: TE pages 316 and 314, CRB page 94, or Transparencies

TEACHING OPTIONS

____ Motivating the Lesson: TE page 317
____ Concept Activity: SE page 315; CRB page 95 (Activity Support Master)
____ Lesson Opener (Visual Approach): CRB page 96 or Transparencies
____ Graphing Calculator Activity with Keystrokes: CRB page 97
____ Examples: Day 1: 1–2, SE pages 316–317; Day 2: 3, SE page 317
____ Extra Examples: Day 1: TE page 317 or Transp.; Day 2: TE page 318 or Transp.; Internet
____ Closure Question: TE page 318
____ Guided Practice: SE page 319; Day 1: Exs. 1–5; Day 2: Exs. 6–10

APPLY/HOMEWORK

Homework Assignment

____ Basic Day 1: 11–24; Day 2: 25–34, 38–43, 44–54 even; Quiz 3: 1–15
____ Average Day 1: 11–24; Day 2: 25–34, 38–43, 44–54 even; Quiz 3: 1–15
____ Advanced Day 1: 11–24; Day 2: 25–43, 44–54 even; Quiz 3: 1–15

Reteaching the Lesson

____ Practice Masters: CRB pages 98–100 (Level A, Level B, Level C)
____ Reteaching with Practice: CRB pages 101–102 or Practice Workbook with Examples
____ Personal Student Tutor

Extending the Lesson

____ Applications (Real-Life): CRB page 104
____ Challenge: SE page 321; CRB page 105 or Internet

ASSESSMENT OPTIONS

____ Checkpoint Exercises: Day 1: TE page 317 or Transp.; Day 2: TE page 318 or Transp.
____ Daily Homework Quiz (5.7): TE page 321 or Transparencies
____ Standardized Test Practice: SE page 321; TE page 321; STP Workbook; Transparencies
____ Quiz (5.6–5.7): SE page 322

Notes ______

LESSON 5.7

TEACHER'S NAME ______________________ CLASS ______ ROOM ______ DATE ______

Lesson Plan for Block Scheduling

1-day lesson (See *Pacing the Chapter,* TE pages 270C–270D) For use with pages 315–322

GOALS
1. Determine whether a linear model is appropriate.
2. Use a linear model to make a real-life prediction.

State/Local Objectives ______________________

____

____

CHAPTER PACING GUIDE	
Day	Lesson
1	5.1 (all); 5.2 (all)
2	5.3 (all)
3	5.4 (all); 5.5 (begin)
4	5.5 (end); 5.6 (begin)
5	5.6 (end); **5.7 (begin)**
6	**5.7 (end)**; Review Ch. 5
7	Assess Ch. 5; 6.1 (all)

✓ Check the items you wish to use for this lesson.

STARTING OPTIONS

____ Homework Check: TE page 311; Answer Transparencies
____ Warm-Up or Daily Homework Quiz: TE pages 316 and 314, CRB page 94, or Transparencies

TEACHING OPTIONS

____ Motivating the Lesson: TE page 317
____ Concept Activity: SE page 315; CRB page 95 (Activity Support Master)
____ Lesson Opener (Visual Approach): CRB page 96 or Transparencies
____ Graphing Calculator Activity with Keystrokes: CRB page 97
____ Examples: Day 5: 1–2, SE pages 316–317; Day 6: 3, SE page 317
____ Extra Examples: Day 5: TE page 317 or Transp.; Day 6: TE page 318 or Transp.; Internet
____ Closure Question: TE page 318
____ Guided Practice: SE page 319; Day 5: Exs. 1–5; Day 6: Exs. 6–10

APPLY/HOMEWORK

Homework Assignment (See also the assignment for Lesson 5.6.)

____ Block Schedule: Day 5: 11–24; Day 6: 25–34, 38–43, 44–54 even; Quiz 3: 1–15

Reteaching the Lesson

____ Practice Masters: CRB pages 98–100 (Level A, Level B, Level C)
____ Reteaching with Practice: CRB pages 101–102 or Practice Workbook with Examples
____ Personal Student Tutor

Extending the Lesson

____ Applications (Real-Life): CRB page 104
____ Challenge: SE page 321; CRB page 105 or Internet

ASSESSMENT OPTIONS

____ Checkpoint Exercises: Day 5: TE page 317 or Transp.; Day 6: TE page 318 or Transp.
____ Daily Homework Quiz (5.7): TE page 321 or Transparencies
____ Standardized Test Practice: SE page 321; TE page 321; STP Workbook; Transparencies
____ Quiz (5.6–5.7): SE page 322

Notes ______________________

____

____

LESSON 5.7

NAME ______________________ DATE __________

Available as a transparency

Warm-Up Exercises

For use before Lesson 5.7, pages 315–322

Write a linear equation for the line through the two points.

1. (21, 14,256) and (45, 39,561)

2. (12.3, 4162) and (14.7, 3219)

Tell whether each scatter plot has positive, negative, or relatively no correlation.

3. the points cannot be approximated by a line

4. the points can be approximated by a line with positive slope

Lesson 5.7

Daily Homework Quiz

For use after Lesson 5.6, pages 307–314

1. Write the equation $y = \frac{7}{10}x - \frac{3}{2}$ in standard form.

2. Write an equation in standard form of the line that passes through $(5, -1)$ and $(-1, 3)$.

3. You are buying apples and oranges for picnic lunches. Apples cost 20¢ apiece and oranges cost 25¢. You have $5 to spend for the fruit. Write an equation in standard form to represent the number of apples x and oranges y you could buy.

NAME ______________________ DATE __________

Activity Support Master

For use with page 315

Step 1

Number of squares in figure	1	2	3	4	5
Number of toothpicks	4				

Step 2

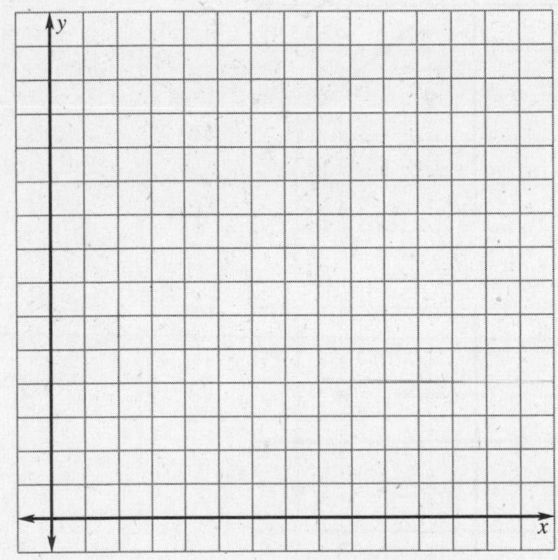

Step 3

Two points: ______________________

Slope: ______________________

Step 4

Equation of the line in slope-intercept form: ______________________

Step 4

Number of toothpicks for 10 squares: ______________________

LESSON

5.7

NAME ______________________ DATE ________

Available as a transparency

Visual Approach Lesson Opener

For use with pages 316–322

Study the graph. Tell whether the line shown is a reasonable model for the data points. Explain your answer.

1.

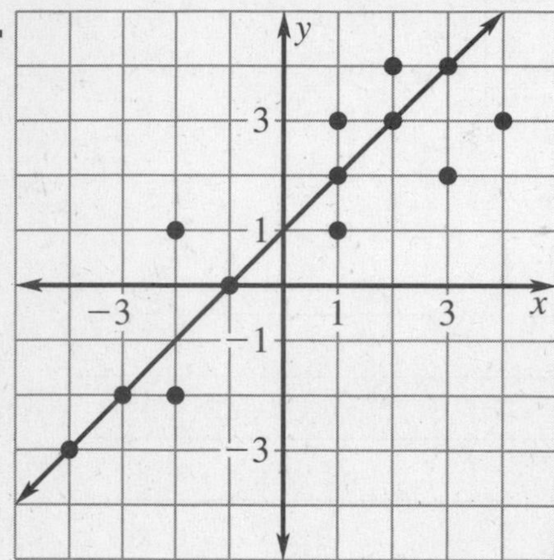

2.

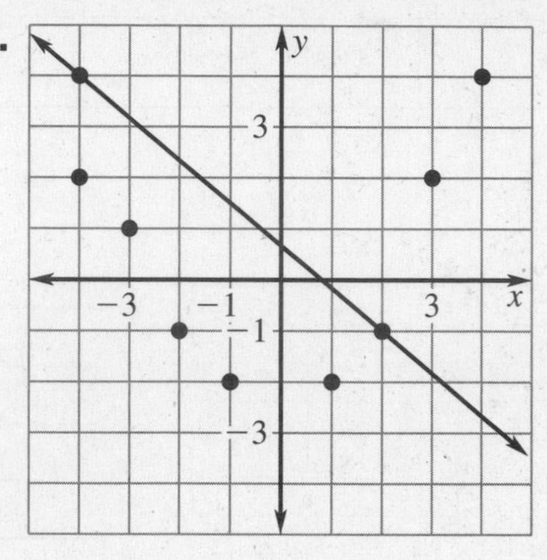

3.

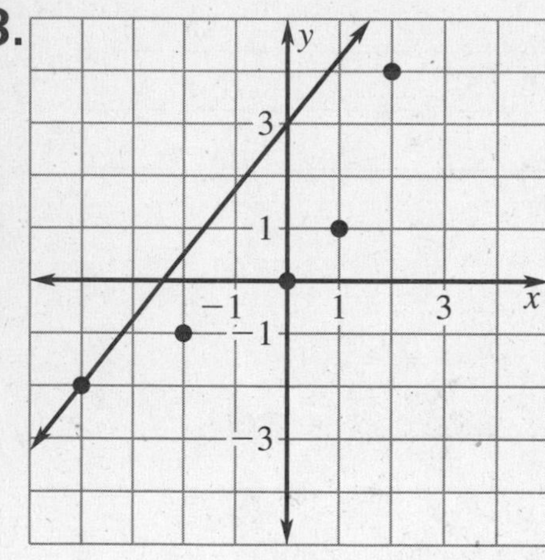

4.

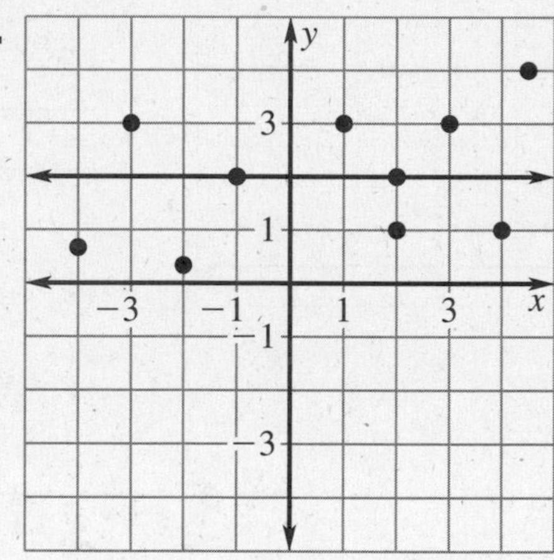

5.

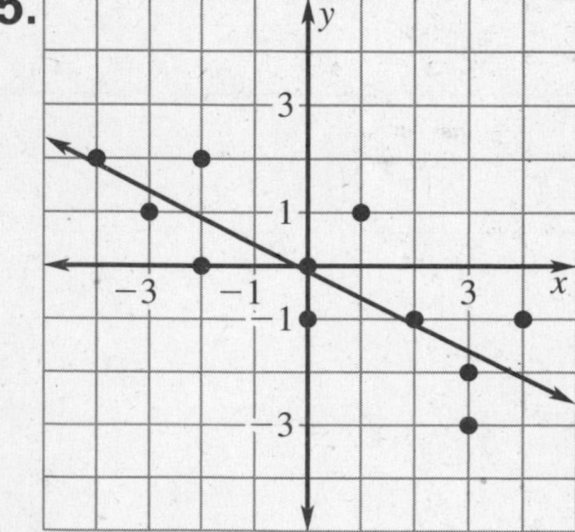

6.

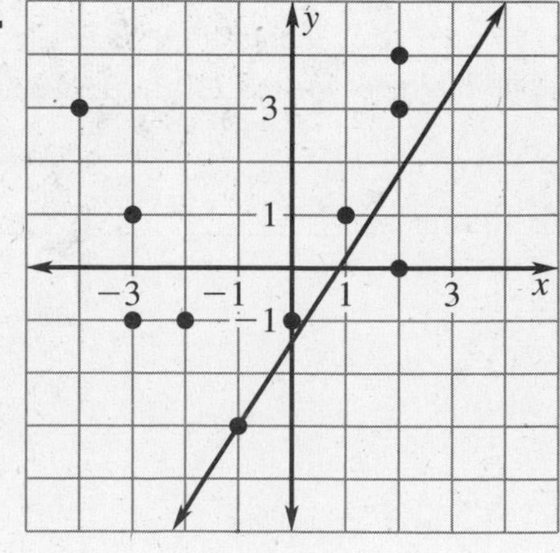

Lesson 5.7

NAME ______________________ DATE __________

Graphing Calculator Activity Keystrokes

For use with page 321

Keystrokes for Exercise 35

TI-82

STAT 1

Enter x-values (years since 1990) in L1.

Enter y-values (average tuition for public colleges) in L2.

2nd [STAT PLOT] 1

Choose the following.

On; Type ⌞⁖⌟; Xlist: L1; Ylist: L2; Mark ▫

Y= 192.64 X,T,θ + 2015.79

WINDOW ENTER (-) 15 ENTER 15 ENTER

3 ENTER (-) 600 ENTER 4200 ENTER

300 ENTER GRAPH 2nd [CALC] 2

Use the cursor keys, ◀ and ▶, to move the trace cursor to select the lower bound at $x \approx -11$. Press ENTER.

Use the cursor keys, ◀ and ▶, to move the trace cursor to select the upper bound at $x \approx -9$. Press ENTER.

Use the cursor keys, ◀ and ▶, to move the trace cursor to select the guess at $x \approx -10$. Press ENTER.

TI-83

STAT 1

Enter x-values (years since 1990) in L1.

Enter y-values (average tuition for public colleges) in L2.

2nd [STAT PLOT] 1

Choose the following.

On; Type ⌞⁖⌟; Xlist: L1; Ylist: L2; Mark ▫

Y= 192.64 X,T,θ,n + 2015.79

WINDOW (-) 15 ENTER 15 ENTER

3 ENTER (-) 600 ENTER 4200 ENTER

300 ENTER GRAPH

2nd [CALC] 2 (-) 11 ENTER (-) 9 ENTER

(-) 10 ENTER

SHARP EL-9600c

STAT [A] ENTER

Enter x-values (years since 1990) in L1.

Enter y-values (average tuition for public colleges) in L2.

2ndF [STAT PLOT] [A] ENTER

Choose the following.

On; Data XY; ListX: L1; ListY: L2

2ndF [STAT PLOT] [G] 3

Y= 192.64 X/θ/T/n + 2015.79

WINDOW (-) 15 ENTER 15 ENTER 3 ENTER

(-) 600 ENTER 4200 ENTER 300 ENTER

GRAPH 2ndF [CALC] 5

CASIO CFX-9850GA PLUS

From the main menu, choose STAT.

Enter x-values (years since 1990) in List 1.

Enter y-values (average tuition for public colleges) in List 2.

F1 F6

Choose the following.

Graph Type: Scatter; XList: List 1; YList: List 2; Frequency: 1; Mark Type: ▫

EXIT

SHIFT F3 (-) 15 EXE 15 EXE 3 EXE (-)

600 EXE 4200 EXE 300 EXE EXIT

F1 F1 OPTN F1 F1 F1 MENU 5

192.64 X,θ,T + 2015.79 EXE

F6 OPTN F1 F2 F1 F5 F1

NAME ______________________ DATE __________

Practice A

For use with pages 316–322

Tell whether it is reasonable for the graph to be represented by a linear model.

1.

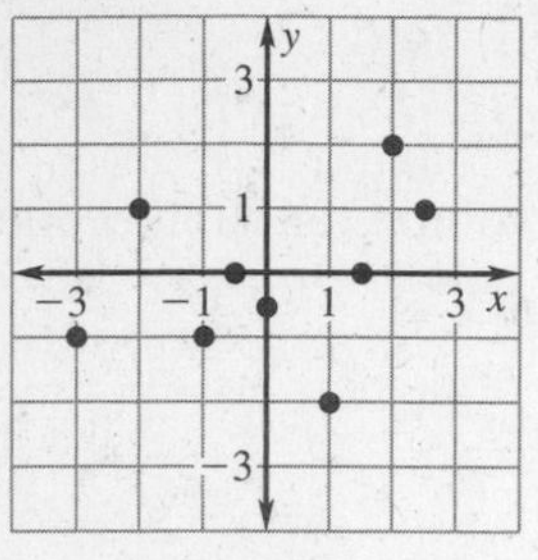

2.

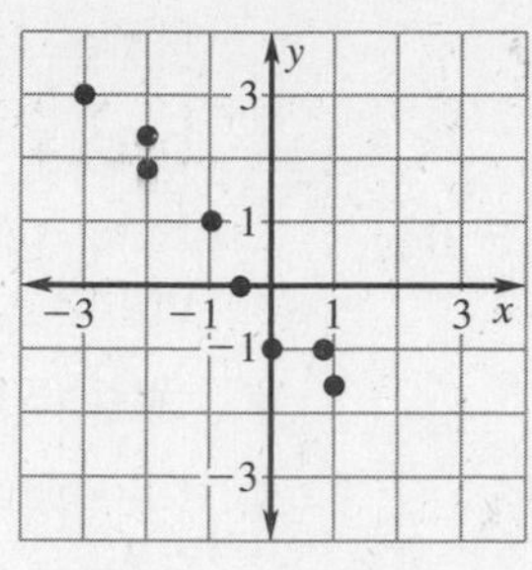

3.

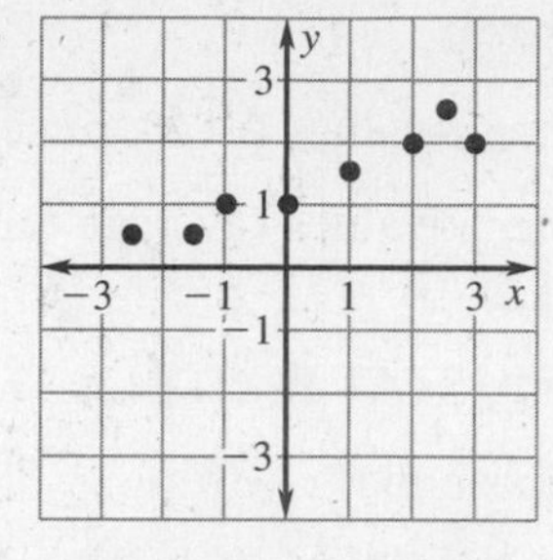

4.

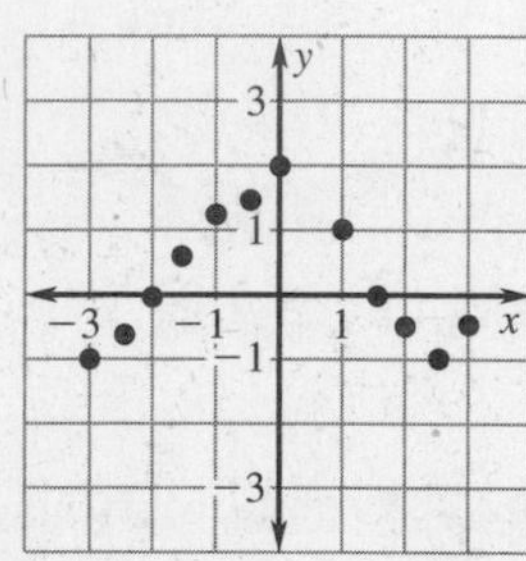

5.

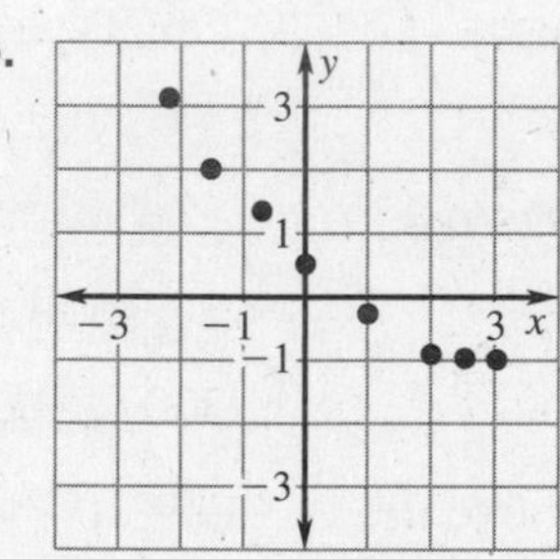

6.

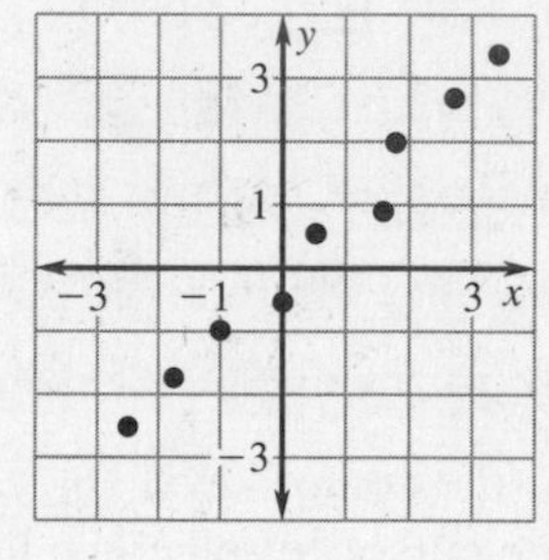

7. Explain the difference between linear interpolation and linear extrapolation.

8. Explain how to decide whether a data set can be represented by a linear model.

Savings Account **In Exercises 9 and 10, use the following information.**

Let $y = 55x + 126$ represent the amount of money (in dollars) in your savings account from 1988 to 1998. Let x represent the number of years since 1988.

9. Use linear interpolation to predict the amount of money in your savings account for 1992.

10. Use linear extrapolation to predict the amount of money in your savings account for 2000.

Movie Prices **In Exercises 11 and 12, use the following information.**

Let $y = 0.25x + 4$ represent the cost of going to a movie from 1985 to 1995. Let x represent the number of years since 1985.

11. Use linear interpolation to predict the cost of going to the movies in 1991.

12. Use linear extrapolation to predict the cost of going to the movies in 1997.

Skim Milk Consumption **In Exercises 13–16, use the table, which shows the number of pounds of skim milk, *S*, consumed per person in the United States in year *t*.**

t	S
1980	26.9
1985	27.4
1990	42.8
1992	46.2
1995	53.9
1996	55.7

13. Make a scatter plot of the pounds of skim milk consumed in terms of the year t. Let t represent the number of years since 1980.

14. Write a linear model for this data.

15. Use the linear model to estimate the number of pounds consumed in 1994.

16. Use the linear model to estimate the number of pounds consumed in 1999.

LESSON **5.7**

NAME ______________________________ DATE ____________

Practice B

For use with pages 316–322

Tell whether it is reasonable for the graph to be represented by a linear model.

1.

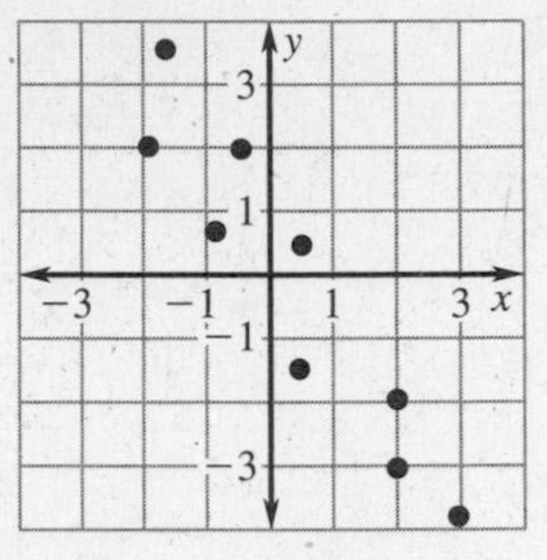

2.

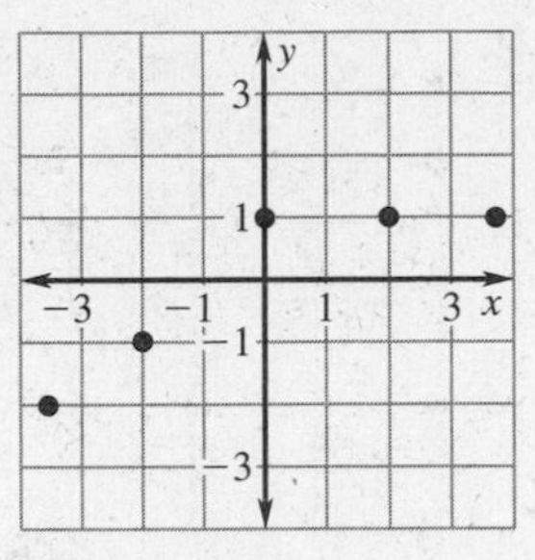

3.

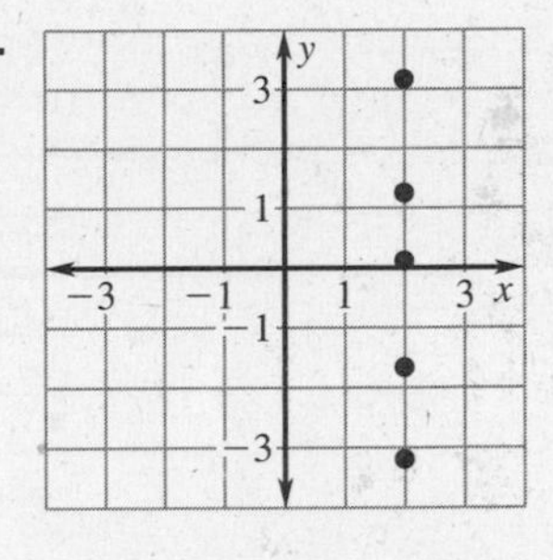

4.

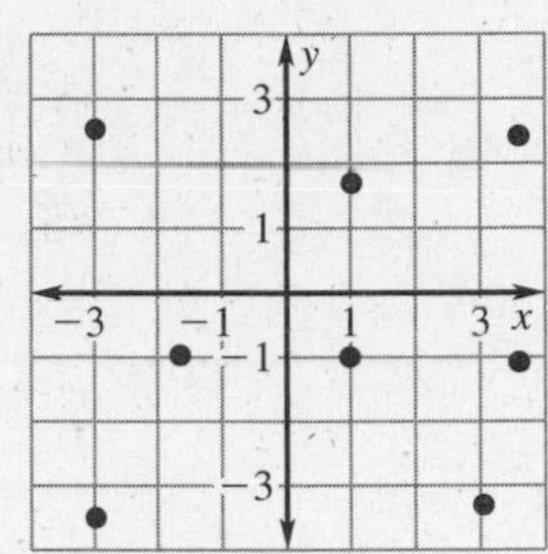

5.

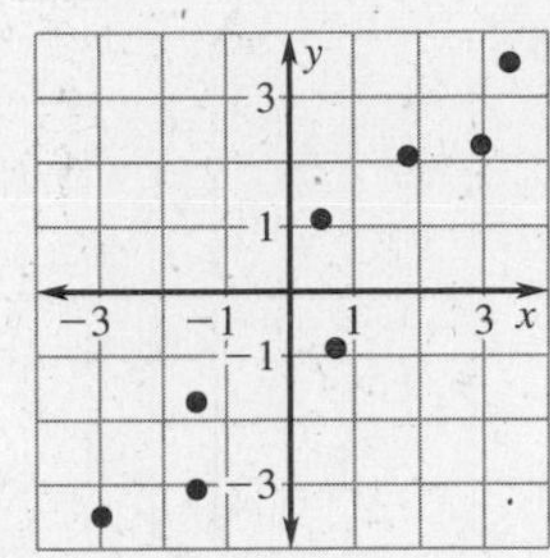

6.

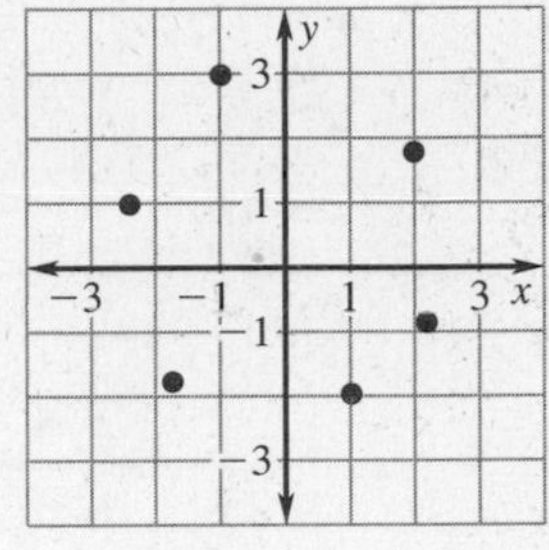

Company Profits **In Exercises 7 and 8, use the following information.**

Let $y = 4.2x + 7.1$ represent a company's profit, in thousands of dollars, from 1985 to 1995. Let x represent the number of years since 1985.

7. Use linear interpolation to predict the profit for 1990.

8. Use linear extrapolation to predict the profit for 1998.

Chemical Workers **In Exercises 9–12, use the table, which shows the average salary, S, of chemical workers in year t.**

t	S
1982	16,731
1985	19,227
1987	21,839
1990	26,269
1992	28,836
1994	30,665
1995	31,345
1997	33,065

9. Make a scatter plot of the average salary of chemical workers in terms of the year t. Let t represent the number of years since 1982.

10. Write a linear model for this data.

11. Use the linear model to estimate the average salary in 1993.

12. Use the linear model to estimate the average salary in 1999.

Whole Milk Consumption **In Exercises 13–16, use the table, which shows the number of pounds of whole milk, W, consumed per person in the United States in year t.**

t	W
1980	141.7
1985	119.7
1990	87.7
1992	81.2
1995	72.6
1996	72.1

13. Make a scatter plot of the pounds of whole milk consumed in terms of the year t. Let t represent the number of years since 1980.

14. Write a linear model for this data.

15. Use the linear model to estimate the number of pounds of whole milk consumed in 1994.

16. Use the linear model to estimate the number of pounds of whole milk consumed in 1999.

NAME ______________________________ DATE ____________

Practice C

For use with pages 316–322

Tell whether it is reasonable for the graph to be represented by a linear model.

1. 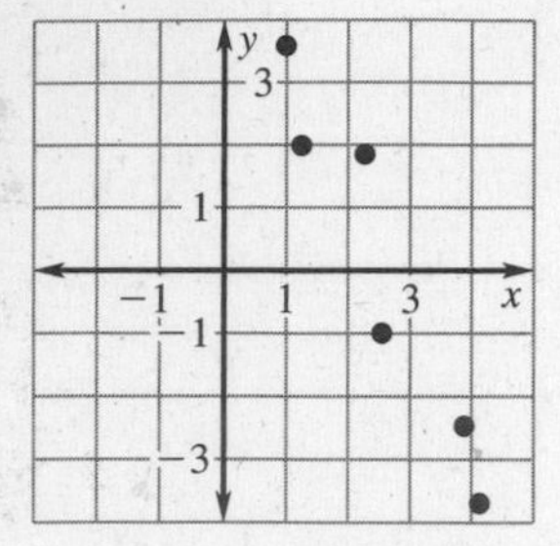

2. 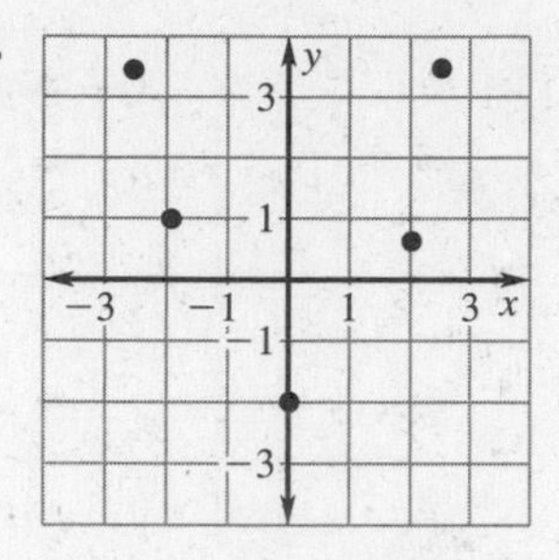

3. 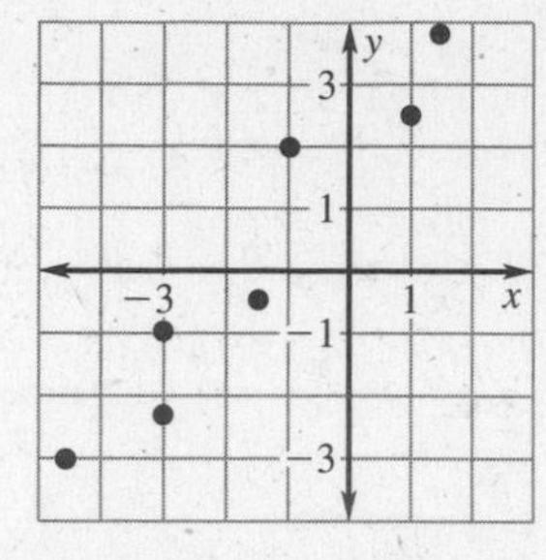

4. 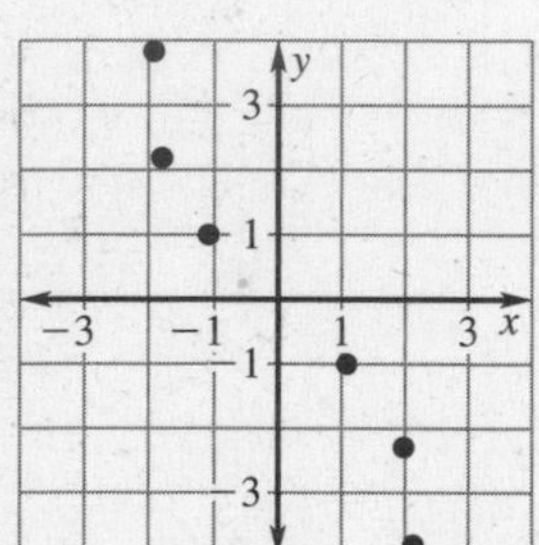

5. 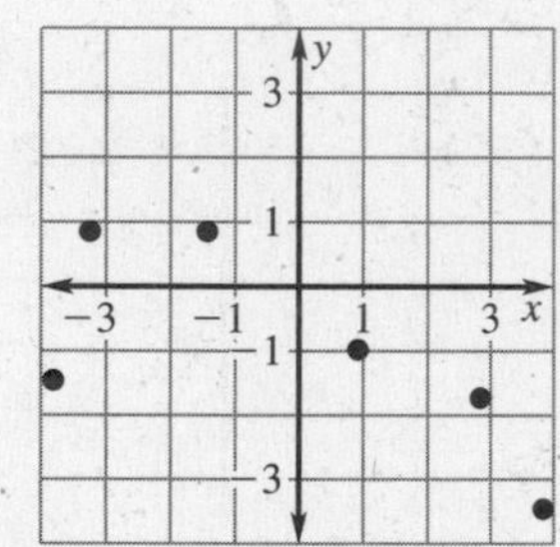

6. 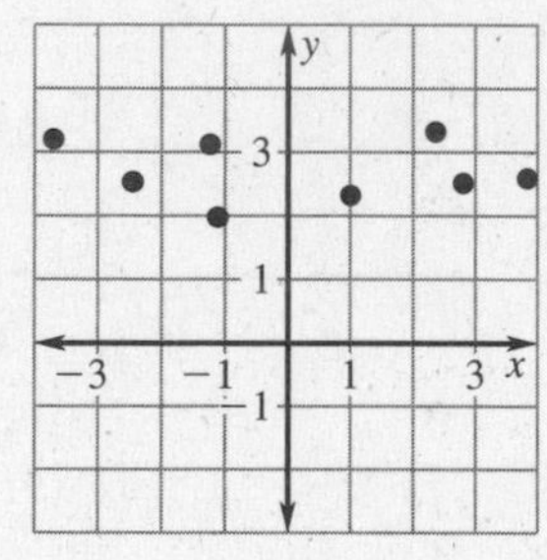

Cost of Advertising **In Exercises 7–9, use the following information.**

For 1980 through 1998, the costs of all local and national advertising (in millions of dollars) can be modeled by the linear equation $y = 5670 + 5655t$, where t represents the number of years since 1980.

7. Use the linear model to predict the total cost of advertising in 2000.
8. Use the linear model to predict the total cost of advertising in 1990.
9. Should the linear model be used to predict the total cost of advertising in 1930? Explain.

School Teachers **In Exercises 10–13, use the table, which shows the average salary, *S*, of public elementary and secondary school teachers in year *t*.**

t	S
1980	15,970
1985	23,600
1990	31,367
1994	35,737
1995	36,685
1996	37,716
1997	38,562

10. Make a scatter plot of the average salary of school teachers in terms of the year t. Let t represent the number of years since 1980.
11. Write a linear model for this data.
12. Use the linear model to estimate the average salary in 1992.
13. Use the linear model to estimate the average salary in 1999.

Density **In Exercises 14–17, use the table, which shows the density factor, *d*, of water at a temperature of *T* (°C).**

T	d
0	28.93
2	28.79
4	28.60
6	28.36
8	28.08

14. Make a scatter plot of the density factor of water in terms of the temperature T.
15. Write a linear model for this data.
16. Use the linear model to estimate the density factor at 7°C.
17. Use the linear model to estimate the density factor at 9°C.

LESSON 5.7

NAME _______________ DATE _______

Reteaching with Practice

For use with pages 316–322

GOAL **Determine whether a linear model is appropriate and use a linear model to make a real-life prediction**

VOCABULARY

Linear interpolation is a method of estimating the coordinates of a point that lies between two given data points.

Linear extrapolation is a method of estimating the coordinates of a point that lies to the right or left of all of the given data points.

EXAMPLE 1 *Is a Linear Model Appropriate?*

Tell whether it is reasonable for the data to be represented by a linear model.

Years since 1995	0	1	2	3
Depreciation (in dollars)	1000	740	520	250

SOLUTION

Draw a scatter plot of the data to decide whether the data can be represented by a linear model. From the scatter plot at the right, you can see that the data fall almost exactly on a line. A linear model is appropriate.

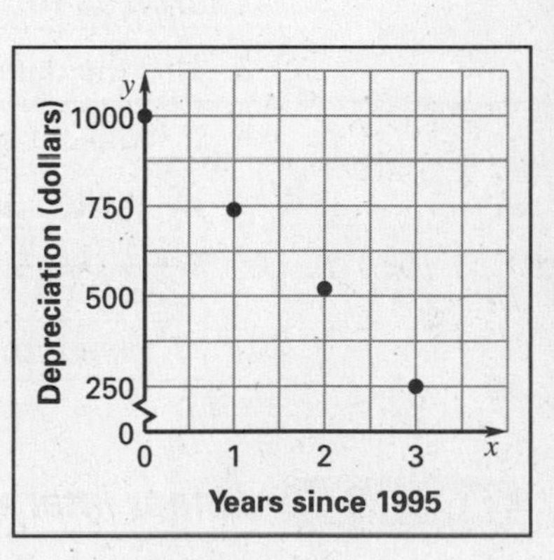

Exercises for Example 1

Tell whether it is reasonable for the graph to be represented by a linear model.

1.

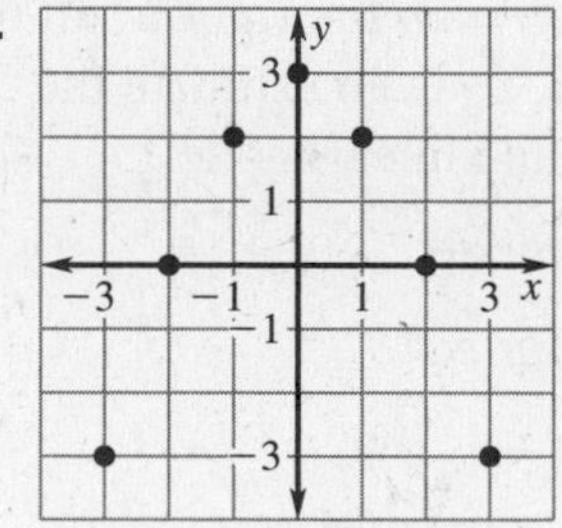

2.

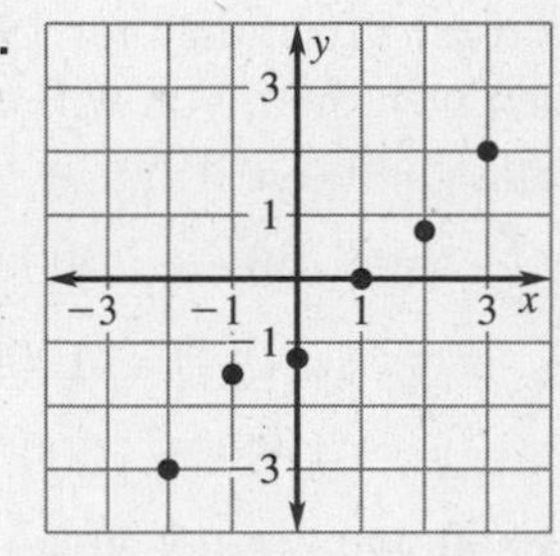

Lesson 5.7

NAME ______________________________ DATE __________

Reteaching with Practice

For use with pages 316–322

EXAMPLE 2 Writing a Linear Model

Use the scatter plot from Example 1 to write a linear model for the data.

SOLUTION

Find two points on the best-fitting line such as (0, 1000) and (3, 250). Use these points to find the slope of the best-fitting line.

$$m = \frac{y_2 - y_1}{x_2 - x_1} = \frac{250 - 1000}{3 - 0} = \frac{-750}{3} = -250$$

Using a y-intercept of $b = 1000$ and a slope of $m = -250$, you can write an equation of the line.

$y = mx + b$ Write slope-intercept form.

$y = -250x + 1000$ Substitute -250 for m and 1000 for b.

A linear model for the data is $y = -250x + 1000$.

Exercises for Example 2

3. Use the data given in the table.

a. Make a scatter plot of the data.

b. Write a linear model for the data.

Year	1990	1992	1994	1996	1998
Expenditures (in millions)	50	210	350	490	650

EXAMPLE 3 Linear Interpolation and Linear Extrapolation

Use the model found in Example 2 to estimate the depreciation in 1999.

SOLUTION

You are given data for 1995–1998. Because 1999 is to the right of all of the given data, you will use linear extrapolation. You can estimate the depreciation in 1999 by substituting $x = 4$ into the linear model.

$y = -250x + 1000$ Write linear model.

$y = -250(4) + 1000$ Substitute 4 for x.

$y = -1000 + 1000 = 0$ Simplify.

The model predicts that the depreciation in the year 1999 will be $0.

Exercise for Example 3

4. Use the model found in Exercise 3 to estimate the expenditures in 1991.

LESSON 5.7

NAME ______________________________ DATE __________

Quick Catch-Up for Absent Students

For use with pages 315–322

The items checked below were covered in class on (date missed) __________

Activity 5.7: Investigating Linear Modeling (p. 315)

____ **Goal:** Decide whether a linear model can represent data.

Lesson 5.7: Predicting with Linear Models

____ **Goal 1:** Determine whether a linear model is appropriate. (p. 316)

Material Covered:

____ Example 1: Which Data Set is More Linear?

____ **Goal 2:** Use a linear model to make a real-life prediction. (pp. 317–318)

Material Covered:

____ Student Help: Look Back

____ Student Help: Study Tip

____ Example 2: Writing a Linear Model

____ Example 3: Linear Interpolation and Linear Extrapolation

Vocabulary:

linear interpolation, p. 318 linear extrapolation, p. 318

____ Other (specify) ______________________________

____

Homework and Additional Learning Support

____ Textbook (specify) pp. 319–322

____

____ Internet: Extra Examples at www.mcdougallittell.com

____ *Reteaching with Practice* worksheet (specify exercises) __________

____ *Personal Student Tutor* for Lesson 5.7

Lesson 5.7

LESSON 5.7

NAME ______________________________ DATE __________

Real-Life Application: When Will I Ever Use This?

For use with pages 316–322

The Internet

The ideas and technology that led to the development of today's Internet first appeard in the 1960s. However, it was about twenty years until the Internet became widely accessible to people outside universities and scientific and government centers.

The Internet allows computers to directly communicate with each other using many kinds of electronic transports including satellite systems, telephone lines, and optical filters. As more people sign onto the Internet, more hosts, or Internet service providers, will be needed. The Internet Software Consortium published the following survey counting the number of Internet hosts.

Year	*Number of Internet Hosts (in Thousands)*
1991	376
1992	727
1993	1,313
1994	2,217
1995	4,852
1996	9,472
1997	16,146
1998	29,670
1999	43,230

1. Make a scatter plot of the data and draw the line that best fits the data.

2. Write a linear model for the number of Internet hosts.

3. Estimate the number of hosts in 2002 using linear extrapolation.

4. Make a scatter plot of the data from 1995 to 1999, and draw the line of best fit.

5. Write a linear model for the number of Internet hosts using the new line of best fit.

6. Estimate the number of hosts in 2002 using the equation from Exercise 5.

NAME ______________________ DATE __________

Challenge: Skills and Applications

For use with pages 316–322

In Exercises 1–2, write a linear model for the data set by finding the *median-median* line as explained in the steps below.

Step 1: Order the data points so that the x-values increase from least to greatest.

Step 2: Group the ordered data into three sets, each containing the same number of points. Find the median of the x's and the median of the y's in each set and write them as ordered pairs (x_1, y_1), (x_2, y_2), (x_3, y_3).

Step 3: Write an equation in the form $y = mx + b$ for the line through (x_1, y_1) and (x_3, y_3).

Step 4: Use your values from Steps 2 and 3 to write an equation of the median-median line $y = mx + \frac{2}{3}b + \frac{1}{3}(y_2 - mx_2)$.

1. $(-8, -28)$, $(-3, -16)$, $(-5, -18)$, $(-1, -7)$, $(5, 12)$, $(0, -6)$, $(3, 4)$, $(7, 15)$, $(2, 5)$
2. $(-3, 16)$, $(-4, 13)$, $(-1, 11)$, $(1, 7)$, $(0, 6)$, $(2, 3)$, $(5, 0)$, $(4, -1)$, $(6, -3)$, $(9, -8)$, $(7, -10)$, $(10, -13)$

In Exercises 3–7, use the table which shows the approximate exchange rate of the Japanese yen per United States dollar in various years. Round decimals in equations to the nearest tenth.

Year	1970	1975	1980	1985	1990	1995
Number of yen per dollar	358	297	227	239	145	94

3. Make a scatter plot of the data and draw the line that best fits the points. Then use two points on your line to write a linear model for the exchange rate between the dollar and the yen t years after 1970.
4. Why is the slope in the model from Exercise 3 negative? What does this mean for the strength of the yen against the dollar?
5. Use the method from Exercises 1 and 2 to write a linear model for the data.
6. Use linear regression on a graphing calculator to write a linear model for the data.
7. Use each model from Exercises 3, 5, and 6 to estimate the exchange rate in 2000. How do your results compare?

CHAPTER 5

NAME ______________________________ DATE ____________

Chapter Review Games and Activities

For use after Chapter 5

Using the given information, write linear equations in SLOPE-INTERCEPT or STANDARD form. The letter associated with the equation in the correct form will answer the riddle when placed on the line with the problem number.

Which trees in the forest get invited to the most parties?

Given information

1. $m = -\frac{6}{5}$ $b = -2$

2. $m = \frac{1}{2}$ $b = 3$

3. $m = \frac{3}{7}$ $b = -7$

4. $(7, -2)$ $m = \frac{8}{7}$

5. $(-4, 1)$ $m = -\frac{13}{6}$

6. $(0, 6)$ $m = \frac{6}{7}$

7. $(-7, 7)$ $(9, -3)$

8. $(1, -1)$ $(4, 5)$

9. $(-6, -4)$ $(-2, -2)$

10. $(9, 0)$ $(-3, 4)$

(S) $x + 3y = 9$
(T) $y = \frac{6}{5}x + 2$
(A) $x + 2y = 6$
(P) $3x - 7y = 49$
(S) $8x - 7y = -42$
(P) $6x + 5y = -10$
(O) $y = \frac{1}{2}x + 3$
(R) $6x - 7y = -42$
(N) $2x - y = 3$
(E) $y = \frac{1}{2}x - 1$
(L) $8x - 7y = 70$
(O) $y = \frac{6}{7}x - 6$
(W) $x + 2y = 2$
(A) $y = -\frac{13}{6}x - 7\frac{2}{3}$
(O) $y = -\frac{5}{8}x + 2\frac{5}{8}$

____ ____ ____ ____ ____ ____   ____ ____ ____ ____
(1) (2) (3) (4) (5) (6)   (7) (8) (9) (10)

Review and Assess

CHAPTER 5

NAME ______________________ DATE ____________

Chapter Test A

For use after Chapter 5

In Questions 1 and 2, write an equation of the line in slope-intercept form.

1. The slope is -5; the y-intercept is 7.
2. The slope is 10; the y-intercept is -3.

Write an equation of the line shown in the graph.

3. 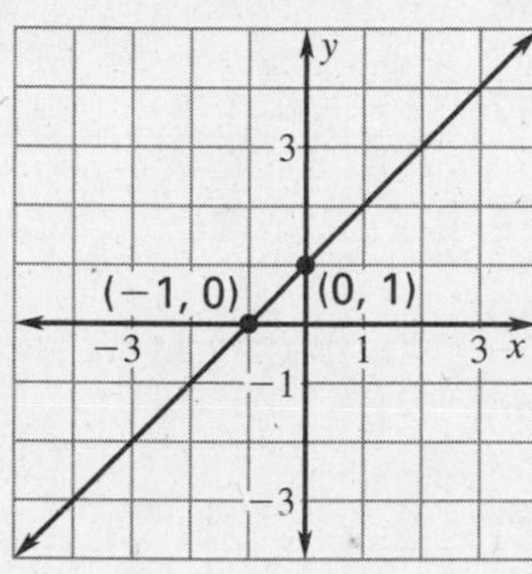

4. 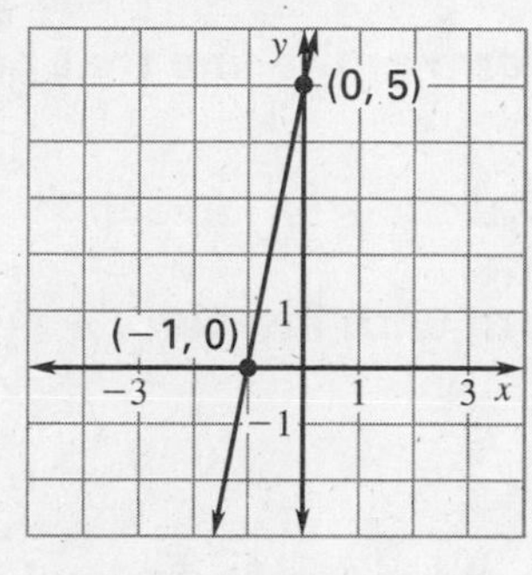

5. Write a linear equation to model the situation. You borrow \$70 from your brother. To repay the loan, you pay him \$7 per week.

Write an equation of the line that passes through the point and has the given slope. Write the equation in slope-intercept form.

6. $(3, 0), m = -2$
7. $(1, 2), m = 2$

Write an equation of the line shown in the graph.

8. 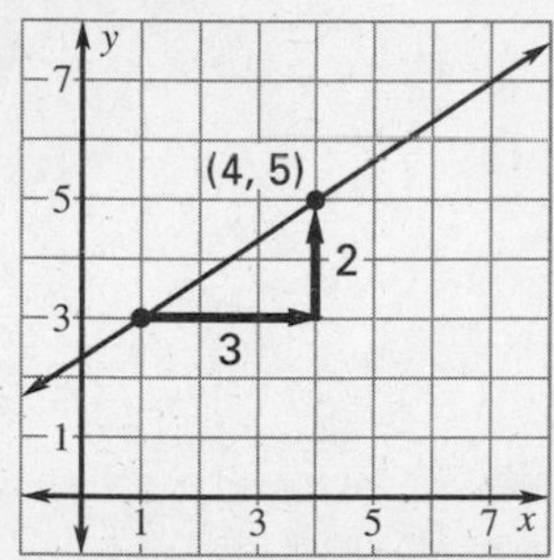

9. 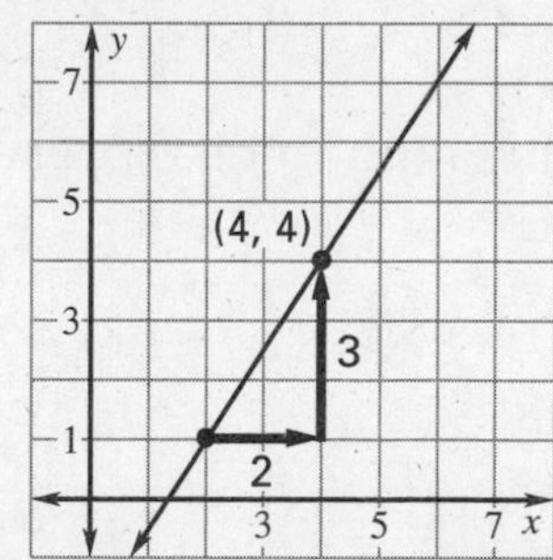

Write an equation of the line that is parallel to the given line and passes through the given point.

10. $y = x + 3, (5, 0)$
11. $y = 2x + 3, (-4, 1)$

Answers

1. ____________
2. ____________
3. ____________
4. ____________
5. ____________
6. ____________
7. ____________
8. ____________
9. ____________
10. ____________
11. ____________

Review and Assess

NAME ______________________ DATE ____________

Chapter Test A

For use after Chapter 5

Write an equation in slope-intercept form of the line that passes through the points.

12. $(-4, 2), (1, -1)$

13. $(-2, -1), (3, 5)$

14. Write an equation of a line that is perpendicular to $y = 2x + 3$ and passes through $(3, 4)$.

Write an equation in point-slope form of the line that passes through the given points.

15. $(-3, -4), (3, 4)$

16. $(-5, -4), (7, -5)$

Write the equation in standard form with integer coefficients.

17. $5x - y + 6 = 0$

18. $y = -3x + 9$

Write the equations in standard form of the horizontal and vertical lines.

19.

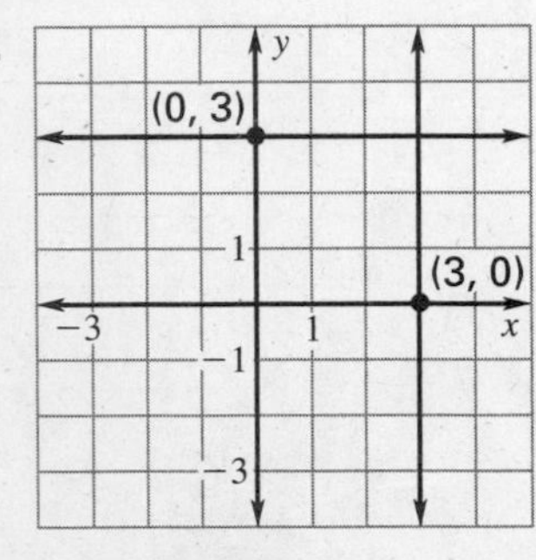

20.

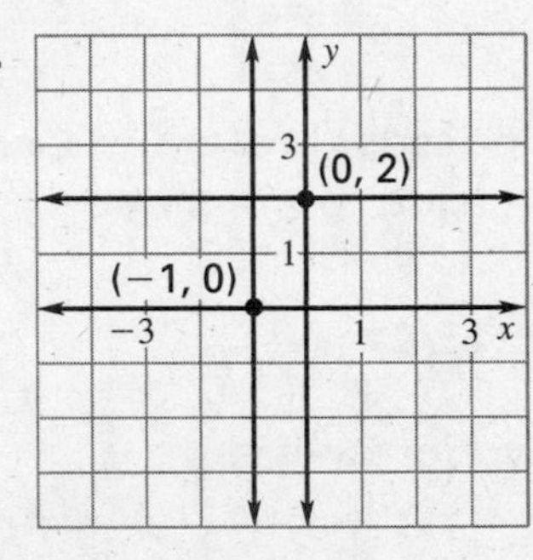

Tell whether it is reasonable for the graph to be represented by a linear model.

21.

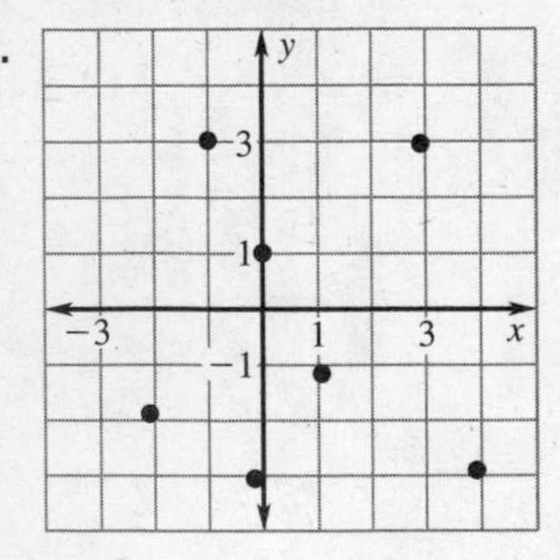

22.

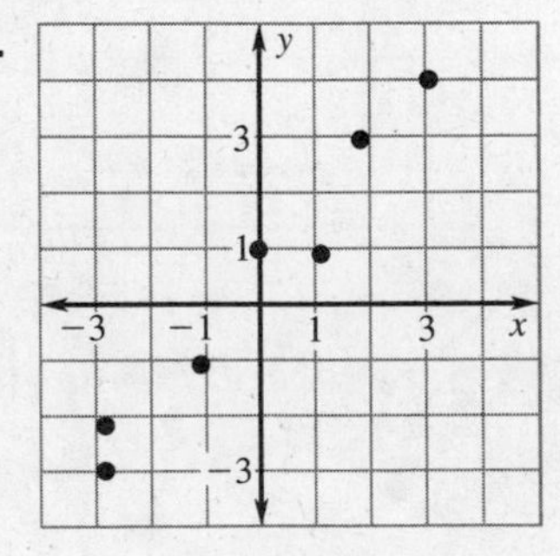

12. ____________
13. ____________
14. ____________
15. ____________
16. ____________
17. ____________
18. ____________
19. ____________
20. ____________
21. ____________
22. ____________

Review and Assess

CHAPTER 5

NAME ______________________ DATE __________

Chapter Test B

For use after Chapter 5

In Questions 1 and 2, write an equation of the line in slope-intercept form.

1. The slope is -3; the y-intercept is 5.
2. The slope is 4; the y-intercept is 0.

Write an equation of the line shown in the graph.

3.

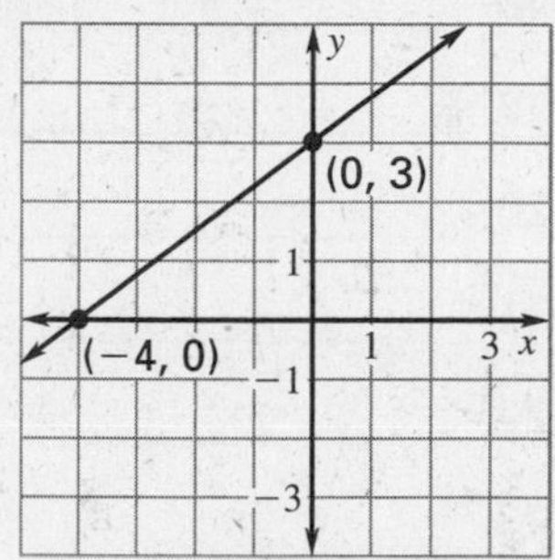

4.

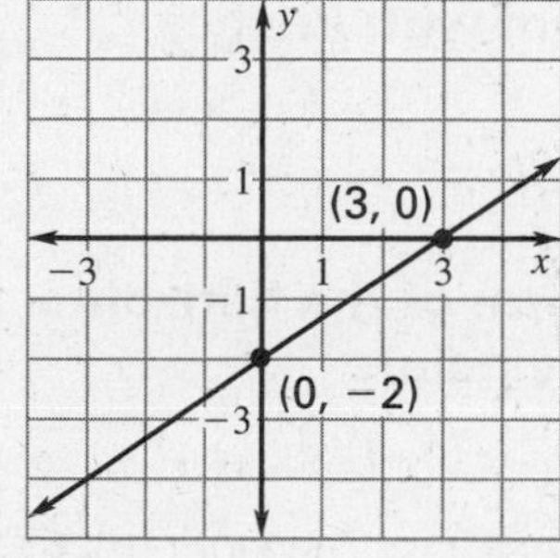

5. Write a linear equation to model the situation. You have walked 4 miles on a trail. You continue to walk at a rate of 3 miles per hour for 5 hours.

Write an equation of the line that passes through the point and has the given slope. Write the equation in slope-intercept form.

6. $(3, 2)$, $m = \frac{1}{2}$

7. $(-3, 2)$, $m = \frac{1}{2}$

Write an equation of the line shown in the graph.

8.

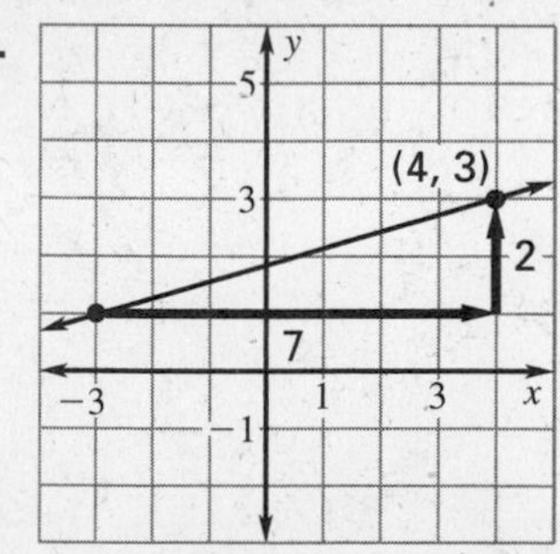

9.

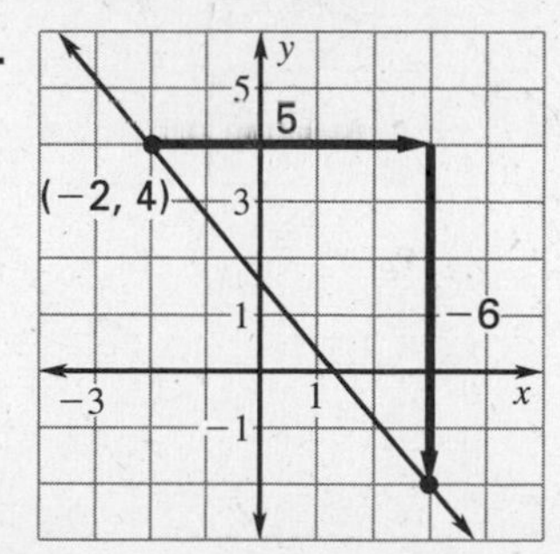

Write an equation of the line that is parallel to the given line and passes through the given point.

10. $y = -3x + 2$, $(2, 3)$

11. $y = \frac{1}{2}x - 5$, $(-3, -1)$

Write an equation in slope-intercept form of the line that passes through the points.

12. $(-5, 3)$, $(4, -5)$

13. $\left(-\frac{1}{2}, -1\right)$, $\left(3, \frac{5}{2}\right)$

Answers

1. ______________
2. ______________
3. ______________
4. ______________
5. ______________
6. ______________
7. ______________
8. ______________
9. ______________
10. ______________
11. ______________
12. ______________
13. ______________

Review and Assess

NAME ______________________________ DATE ____________

Chapter Test B

For use after Chapter 5

14. Write an equation of a line that is perpendicular to $y = -3x + 5$ and passes through $(4, 3)$.

Write an equation in point-slope form of the line that passes through the given points.

15. $(5, -6), (1, -7)$

16. $(6, -3), (-1, 9)$

Write the equation in standard form with integer coefficients.

17. $0.5x - 2y - 0.75 = 0$

18. $y = -\frac{1}{3}x - 5$

Write the equations in standard form of the horizontal and vertical lines that pass through the point.

19. $(2, 4)$

20. $(-5, 4)$

21. You are at the music store looking for CDs. The store has CDs for \$10 and \$15. You have \$55.00 to spend. Write an equation that represents the different numbers of \$10 and \$15 CDs that you can buy.

Tell whether it is reasonable for the graph to be represented by a linear model.

22.

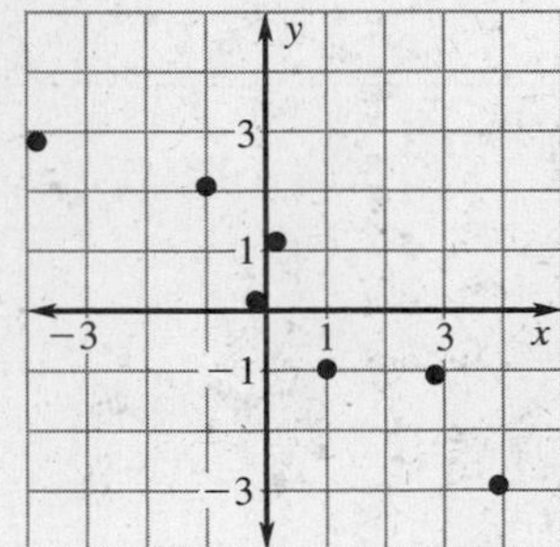

23.

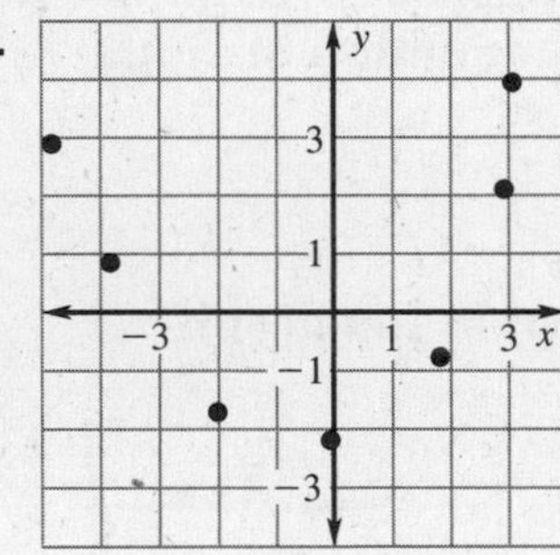

14. ____________

15. ____________

16. ____________

17. ____________

18. ____________

19. ____________

20. ____________

21. ____________

22. ____________

23. ____________

CHAPTER 5

NAME ______________________________ DATE ____________

Chapter Test C

For use after Chapter 5

In Questions 1 and 2, write an equation of the line in slope-intercept form.

1. The slope is $-\frac{4}{3}$; the y-intercept is -2.
2. The slope is 0; the y-intercept is -5.
3. You are driving at a speed of 55 miles per hour to your sister's house. At 1 P.M. you are 150 miles from her house. Write an equation that models the distance from your sister's house in terms of the number of hours since 1 P.M.

Write an equation of the line that passes through the point and has the given slope. Write the equation in slope-intercept form.

4. $(4, -5)$, $m = \frac{2}{3}$
5. $(-4, 3)$, $m = -\frac{3}{4}$

Write an equation of the line shown in the graph.

6.

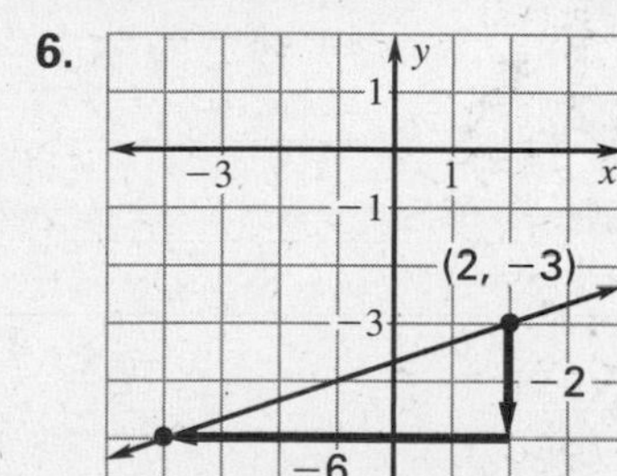

7.

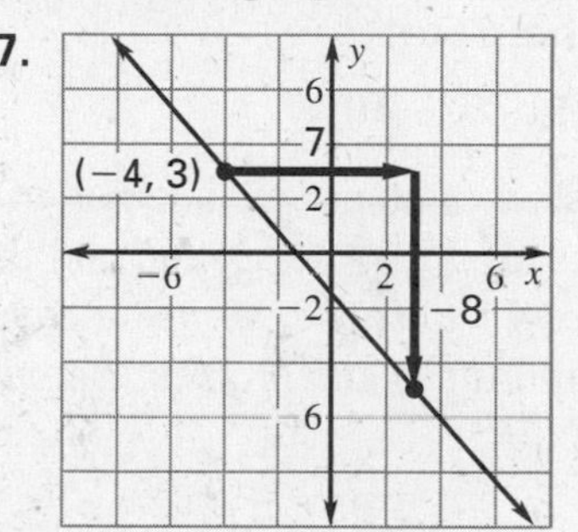

Write an equation of the line that is parallel to the given line and passes through the given point.

8. $y = -4x - 7$, $(5, -3)$
9. $y = -\frac{2}{3}x + 4$, $(-5, 5)$
10. Water freezes at 32° Fahrenheit (or 0° Celsius) and boils at 212° Fahrenheit (or 100° Celsius). Write a linear equation that models the temperature in degrees Fahrenheit F in terms of the temperature in degrees Celsius C.

Answers

1. ____________
2. ____________
3. ____________
4. ____________
5. ____________
6. ____________
7. ____________
8. ____________
9. ____________
10. ____________

NAME ______________________ DATE ________

Chapter Test C

For use after Chapter 5

Write an equation in slope-intercept form of the line that passes through the points.

11. $(-6, 1), (3, -7)$

12. $\left(-\frac{2}{3}, 4\right), \left(6, -\frac{1}{3}\right)$

13. Write an equation of a line that is perpendicular to $y = -\frac{3}{4}x + 2$ and passes through $(-4, 1)$.

Write an equation in point-slope form of the line that passes through the given points.

14. $(-2, 5), (4, -3)$

15. $\left(\frac{1}{2}, -1\right), \left(-\frac{2}{3}, 6\right)$

Write the equation in standard form with integer coefficients.

16. $y = \frac{1}{2}x - \frac{3}{5}$

17. $y = -\frac{8}{3}x + \frac{4}{9}$

Write the equations in standard form of the horizontal and vertical lines that pass through the point.

18. $(-5, -7)$

19. $(6, -10)$

20. You are in charge of bringing snacks to a picnic. You decide to buy grapes and pretzels. The grapes are $2.39 per pound and the pretzels cost $1.89 a bag. You have $11 to spend. Write an equation that models the different amounts of grapes and pretzels that you can buy.

11. ____________
12. ____________
13. ____________
14. ____________
15. ____________
16. ____________
17. ____________
18. ____________
19. ____________
20. ____________
21. ____________
22. ____________
23. ____________

In Questions 21–23, use the following information. The table shows the yearly profits for a company.

Year	1994	1995	1996	1997	1998
Profit (in thousands of dollars)	$250,000	$280,000	$240,000	$320,000	$310,000

21. Make a scatter plot and fit a line to the data.

22. Write a linear model for the amount of profit.

23. Use the linear model to estimate the profit in 2002.

CHAPTER 5

NAME ______________________________ DATE __________

SAT/ACT Chapter Test

For use after Chapter 5

1. What is the equation of the line that passes through the points $(-3, 4)$ and $(-9, 6)$?

   Ⓐ $y = -\frac{1}{3}x - \frac{5}{3}$ Ⓑ $y = -\frac{1}{3}x + 3$
   Ⓒ $y = -3x - 5$ Ⓓ $y = -3x + 12$

2. A line with a slope of -3 passes through the point $(4, -3)$. If $(-3, p)$ is another point on the line, what is the value of p?

   Ⓐ -21 Ⓑ 0
   Ⓒ 18 Ⓓ 24

3. An equation of the line parallel to the line $y = \frac{1}{3}x - 2$ and passes through $(3, -5)$ is __?__.

   Ⓐ $y = -3x + 4$ Ⓑ $y = \frac{1}{3}x + \frac{14}{3}$
   Ⓒ $y = -3x - 12$ Ⓓ $y = \frac{1}{3}x - 6$

4. An equation of the line perpendicular to the line $y = -\frac{3}{4}x + 4$ with a y-intercept of -5 is __?__.

   Ⓐ $y = -\frac{3}{4}x - 5$ Ⓑ $y = \frac{3}{4}x - 5$
   Ⓒ $y = \frac{4}{3}x - 5$ Ⓓ $y = -\frac{4}{3}x + 5$

5. What is the equation of the line that passes through $(-6, 2)$ and has a slope of $-\frac{2}{3}$?

   Ⓐ $y = -\frac{2}{3}x - \frac{14}{3}$ Ⓑ $y = -\frac{2}{3}x - 2$
   Ⓒ $y = -\frac{2}{3}x + 6$ Ⓓ $y = -\frac{2}{3}x - 6$

6. What is the equation of the line shown in the graph?

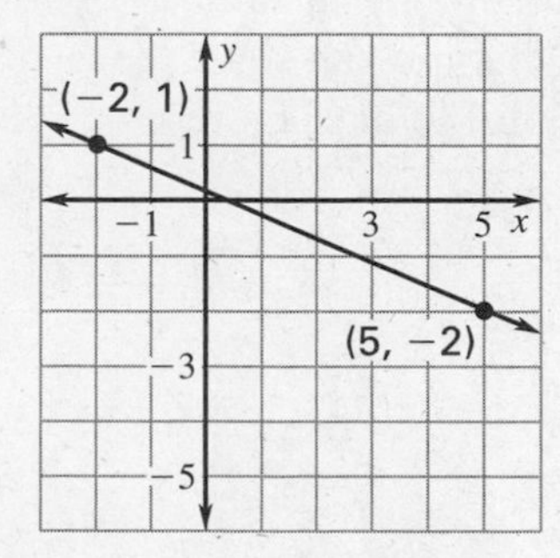

   Ⓐ $y = -x - 1$
   Ⓑ $y = -\frac{3}{7}x + 3$
   Ⓒ $y = -x + 3$
   Ⓓ $y = -\frac{3}{7}x + \frac{1}{7}$

In Questions 7 and 8, choose the statement below that is true about the given numbers.

A. The number in column A is greater.
B. The number in column B is greater.
C. The two numbers are equal.
D. The relationship cannot be determined from the given information.

7.

Column A	*Column B*
x-intercept of $2x - 3y = 4$	x-intercept of $3x - 7y = -6$

Ⓐ Ⓑ Ⓒ Ⓓ

8.

Column A	*Column B*
slope of $9x - 12y = 8$	slope of $4y - 3x = 16$

Ⓐ Ⓑ Ⓒ Ⓓ

9. What is the equation of the line that best fits the scatter plot?

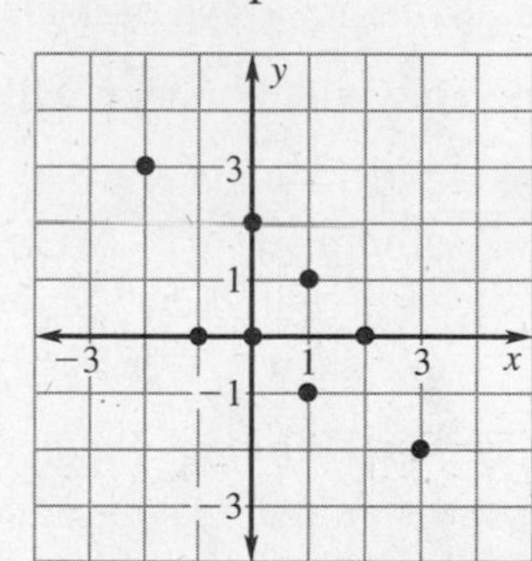

   Ⓐ $y = -x + 1$
   Ⓑ $y = -x - 1$
   Ⓒ $y = x + 1$
   Ⓓ $y = x - 1$

Review and Assess

CHAPTER 5

NAME ______________________ DATE __________

Alternative Assessment and Math Journal

For use after Chapter 5

JOURNAL 1. In Chapter Five we have been studying writing equations of lines including writing equations of best-fit lines from a scatter plot. Why are best fit lines important? In this journal we will explore the use of data in the real world. (a) Use the data set (0, 2), (1, 4), (2, 7), (3, 8), (4, 10), and (5, 12). Make a scatter plot of the data. (b) Write the equation of the best-fit line for the data. Graph the best-fit line on the scatter plot. Explain how to find the best-fit line and what purpose it serves. (c) Write a paragraph explaining the data trends. Create a possible scenario to explain the trends if $x = 0$ represents 1990. (d) Make a prediction for future outcomes using your equation. Write and supply the answer to a question about a future value of the data. Clearly explain your answer and state whether or not it is reasonable.

MULTI-STEP PROBLEM 2. Zachary and Jeremy both have savings accounts at the local bank. Jeremy has a job and has been saving his money. Zachary relies on his account for spending money.

- Jeremy had $500 in his account at the end of 3 weeks, and $800 in his account at the end of 7 weeks.
- Zachary withdraws $7.50 per week from his account. After twelve weeks, the balance in his account was $432.50.

a. Find the rate of change in both Zachary and Jeremy's accounts. Describe what each means in terms of the particular account.

b. Write a model for each which gives the balance of each account, y, in terms of the number of weeks, x.

c. Sketch a graph of each model and label your axes.

d. How much money will be in each account after 20 weeks?

e. After how many weeks will Zachary run out of money? Can he make a withdrawal of $7.50 in the final week? If yes, explain. If no, how much can he withdraw? Show your work and give an explanation to support your answer.

3. ***Critical Thinking*** Predict when Jeremy and Zachary will have the same amount in their savings accounts. Show three different methods that can be used to solve this problem. Show all work and explain why you can solve the problem using these methods. Explain why it can be helpful to have more than one method for solving a particular problem.

Review and Assess

Alternative Assessment Rubric

For use after Chapter 5

JOURNAL SOLUTION

1. Explain that best-fit lines are important as they can be used to predict future data points and they help to analyze the trends of data.

   a–d. Complete answers should address these points:

   a. Check graphs.

   b. Find an equation of the best-fit line, for example, $y = 1.97x + 2.24$. For a more accurate approach students should use a graphing calculator to find the regression equation.

   Explain a method (either selecting points or using a calculator). Explain that a linear model can be used to estimate data points that are not given.

   c. Explain data is increasing over time.

   Create a scenario to explain data trends. *Sample answer:* A company's sales from 1990 to 1995, in millions of dollars. Each year the company is growing, and sales are increasing.

   d. Choose an additional x-value to substitute into the best-fit line equation to determine a future y-value. *Sample answer:* In 1996, $x = 6$, the company would make about \$14 million.

   Explain whether the answer is reasonable based on their specific data.

MULTI-STEP PROBLEM SOLUTION

2. **a.** 75; -7.5; Jeremy deposits \$75 weekly; Zachary withdraws \$7.50 weekly.

   b. Jeremy: $y = 75x + 275$; Zachary: $y = -7.50x + 522.50$

   c. Check graphs.

   d. Jeremy: \$1775; Zachary: \$372.50

   e. 70 weeks; No; He can withdraw \$5 in the last week.

3. ***Critical Thinking Solution*** They will have approximately the same amount after 3 weeks. *Sample answer:* One method would be to set the two equations equal to each other, $75x + 275 = -7.50x + 522.50$, and solve for x. Another method would be to sketch the graph of both equations and find out where the two equations have the same x-value. This would occur where the two equations intersect. A third method would be to use a table of values, either on the calculator or by hand, and calculate the balances for several weeks to determine the solution.

MULTI-STEP PROBLEM RUBRIC

4 Students complete all parts of the questions accurately. Explanations are logical. Two (or three using extension) different methods are clearly explained for determining when the accounts would be approximately equal. Equations and graphs are completely correct.

3 Students complete the questions and explanations. Solutions may contain minor mathematical errors, but errors are carried through (i.e. if the incorrect equations are given, these incorrect equations are graphed accurately). Graphs are correct. Some explanations may not be completely clear.

2 Students complete questions and explanations. Several mathematical errors may occur. Explanations are not logical. Graph is incomplete or incorrect.

1 Answers are very incomplete. Solutions and reasoning are incorrect. Graph is missing or is completely inaccurate. Explanations are not logical.

NAME ______________________ DATE ____________

Project: What We Read

For use with Chapter 5

OBJECTIVE **Analyze the amounts spent by United States consumers on books and maps as opposed to magazines, newspapers, and sheet music.**

MATERIALS paper, pencil, calculator, graph paper

INVESTIGATION The Bureau of Economic Analysis, United States Department of Commerce, groups reading materials as shown in the table.

Personal Consumer Purchases in the United States (in billions of dollars)

	1990	*1992*	*1994*	*1996*
Books and maps	17.5	17.7	20.6	23.2
Magazines, newspapers, and sheet music	23.8	21.6	24.5	26.5

1. Use scatter plots to determine which data are better modeled with a linear model.
2. Write a linear model for the amount spent on books and maps. Write a linear model for the amount spent on magazines, newspapers, and sheet music. For each, let t be years since 1990 and y be amount spent.
3. Use the linear models from Exercise 2 to estimate the amount of purchases of each type in 1993. Did you use *linear interpolation* or *linear extrapolation*?
4. In 1993, people in the United States spent $22.1 billion on books and maps and $22.7 billion on magazines, newspapers, and sheet music. Which of your estimates from Exercise 3 was closer to the actual amount?
5. Use the linear models from Exercise 2 to estimate the amount of purchases of each type in 2000. Did you use *linear interpolation* or *linear extrapolation*?
6. Find the amount spent on each type of reading material in the year 2000 according to the Bureau of Economic Analysis. Which of your estimates from Exercise 5 was closer to the actual amount?

PRESENT YOUR RESULTS Analyze which linear model fits the data better overall. Write a report about your analysis. Include your equations, estimates, and predictions. Discuss which model is best for interpolation and which is best for extrapolation.

Project: Teacher's Notes

For use with Chapter 5

GOALS

- Write an equation of a line given two points on the line.
- Use a linear equation to model a real-life problem.
- Find a linear equation that approximates a set of data points.
- Determine whether a linear model is appropriate.
- Use a linear model to make a real-life prediction.

MANAGING THE PROJECT You may wish to have students write the equations and make predictions before doing research in Exercise 6. After the project, you may wish to pair up students who used different data points to form their models. Have them compare their results and discuss why they chose the points they did.

RUBRIC The following rubric can be used to assess student work.

4 The student finds a linear model for the amount spent on books and maps and for the amount spent on magazines, newspapers, and sheet music, uses the models to estimate and predict, and analyzes the fit of the models. The report demonstrates clear thinking and explanation. All work is complete and correct.

3 The student finds a linear model for the amount spent on books and maps and for the amount spent on magazines, newspapers, and sheet music, uses the models to estimate and predict, and analyzes the fit of the models. However, the report indicates some minor misunderstanding of content, there are errors in computation, or the presentation is weak.

2 The student partially achieves the mathematical and project goals of finding a linear model for the amount spent on books and maps and for the amount spent on magazines, newspapers, and sheet music, using the models to estimate and predict, and analyzing the fit of the models. However, the report indicates a limited grasp of the main ideas or requirements. Some of the work is incomplete, misdirected, or unclear.

1 The student makes little progress toward accomplishing the goals of the project because of lack of understanding or lack of effort.

CHAPTER 5

NAME ______________________ DATE __________

Cumulative Review

For use after Chapters 1–5

Evaluate the expression for the given values of the variable. (1.2)

1. $(2s + t)^3$ when $s = -1$ and $t = -5$
2. $a - \frac{1}{2}b^3$ when $a = \frac{1}{3}$ and $b = -\frac{2}{3}$
3. $-10.9(k - 0.36)$ when $k = 0.45$
4. $8 - (x^2) - (y^3)$ when $x = -3$ and $y = 6$

Write an equation or an inequality to model the situation. (1.5)

5. The length l of a table is three times its width w.
6. The time t it takes to go to work from home is less than one half the time s it takes to go to the mall.
7. The height h of a triangle is equal to the quotient of two times the area a and its base b.

Find the sum or difference. (2.2–2.3)

8. $|-3| - |2^2| - (2)$
9. $-23.8 - 0.327 + 45.96$
10. $|-3.677| + 4.279 - 5.698$
11. $-\frac{7}{3} - |-\frac{5}{6}| + \left(\frac{2}{3}\right)^2$

Solve the equation. (3.1–3.4)

12. $-14x - 5 = 93$
13. $-15a + 30 = -89$
14. $-(x - 2.3) - 4(5.9 - x) = 80$
15. $\frac{2}{9}\left(x - \frac{6}{7}\right) = \left(\frac{1}{63}x + \frac{40}{63}\right)$

Find the *x*- and *y*-intercepts of the equation. (4.3)

16. $y = 4x - 14$
17. $\frac{8}{9}x + \frac{40}{27} = 10y$

Find the slope of the line passing through the given points. (4.4)

18. $(0, -18), (-3.5, -18.99)$
19. $\left(\frac{8}{3}, \frac{4}{7}\right), \left(\frac{50}{3}, \frac{6}{21}\right)$

Write an equation of the line with the given slope and *y*-intercept. (5.1)

20. $m = 3, b = -2$
21. $m = -5, b = 4$
22. $m = 7, b = 11$
23. $m = \frac{1}{5}, b = -6$
24. $m = 0, b = 8$
25. $m = -6.5, b = 4.5$

Write an equation of the line that is parallel to the given line and passes through the given point. (5.2)

26. $y = 4x + 6, (4, -2)$
27. $y = -\frac{1}{4}x - 1, (4, 1)$
28. $y = -3x + 9, (3, -2)$
29. $y = \frac{4}{3}x - 6, (3, 1)$

Review and Assess

CHAPTER 5 CONTINUED

NAME ______________________ DATE __________

Cumulative Review

For use after Chapters 1–5

Write an equation in slope-intercept form of the line that passes through the given points. (5.3)

30. $(12, -3), (-8, 1)$

31. $(-12, -56), (-40, 0)$

32. $\left(\frac{1}{2}, -\frac{1}{2}\right), \left(\frac{1}{8}, \frac{5}{8}\right)$

33. $(-3.75, 3), (3.23, -3.44)$

For Exercises 34–36: (5.4)

a) Make a scatter plot of the data.

b) State whether *x* and *y* have a positive correlation, negative correlation, or relatively no correlation.

c) If possible, write an equation of a best fitting line for the scatter plot.

34.

x	y
0	-6
0.5	2
5	7
8	2

35.

x	y
$-1\frac{5}{6}$	2
0	$1\frac{1}{2}$
$\frac{1}{2}$	0
$2\frac{3}{4}$	-4

36.

x	y
1.2	-8
2.3	-6
4.1	-4
5.6	-3.1

Write an equation in point-slope form of the line that passes through the given points. (5.5)

37. $(2, 7), (-2, -7)$

38. $(-10, 8), (-20, -12)$

39. $(4, 1), (-2, -3)$

40. $(-10, 10), (-6, -6)$

Write the standard form of the equation of the line passing through the given point that has the given slope. (5.6)

41. $(2, -5), m = -5$

42. $(0, 5), m = -\frac{1}{2}$

43. $(-2, 7.5), m = -6$

44. $(-4, 4), m = 4$

In Exercises 45–47, use the following information regarding the farm population (millions of persons) from 1945 to 1995. (5.7)

Year	1945	1955	1965	1975	1985	1995
Millions of Persons	24.3	19.0	12.3	8.8	2.0	0.04

45. What was the average farm population in 1975?

46. Make a scatter plot of the number of people living on farms in terms of the year x. Let $x = 0$ represent 1945.

47. Write a linear model for the farm population (millions of persons).

Review and Ass

ANSWERS

Chapter Support

Parent Guide

Chapter 5

5.1: $y = 0.18t + 15$; \$18.60 **5.2:** $y = \frac{1}{2}x + 9$
5.3: $y = -2x + 5$ **5.4:** negative correlation; the value of a dollar has been decreasing as the years since 1900 have increased.
5.5: $y + 4 = 4(x + 5)$ or $y - 8 = 4(x + 2)$
5.6: $10x + 15y = 60$; 6 sweatshirts and no jeans, no sweatshirts and 4 jeans, 3 sweatshirts and 2 jeans **5.7:** *Sample answer:* $y = 0.563x + 2.43$; about \$19.3 million

Prerequisite Skills Review

1. -2 **2.** no solution **3.** 5 **4.** $\frac{1}{2}$

5.

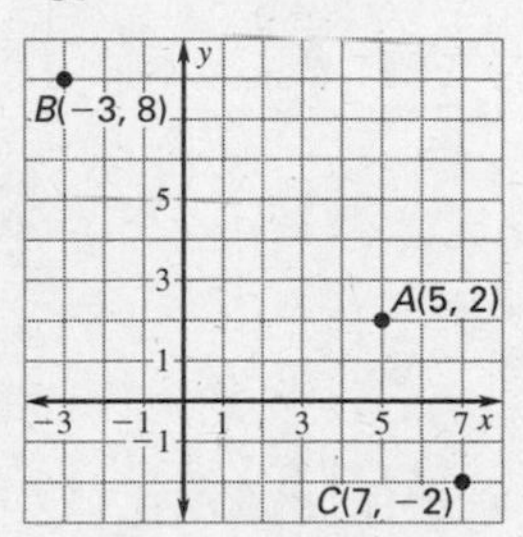

6.

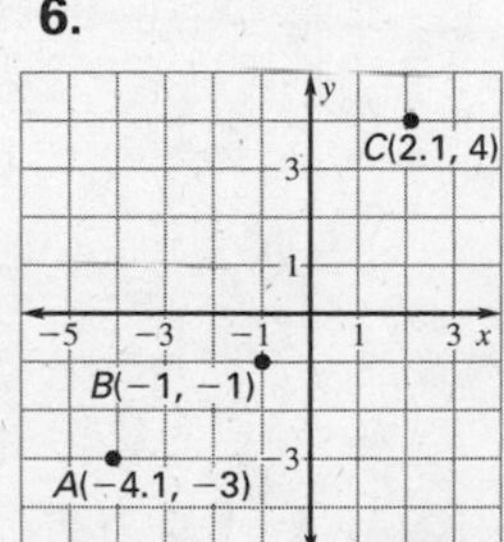

7.

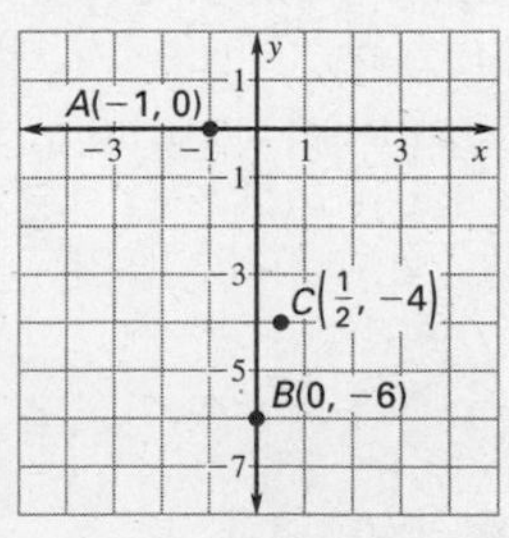

8.

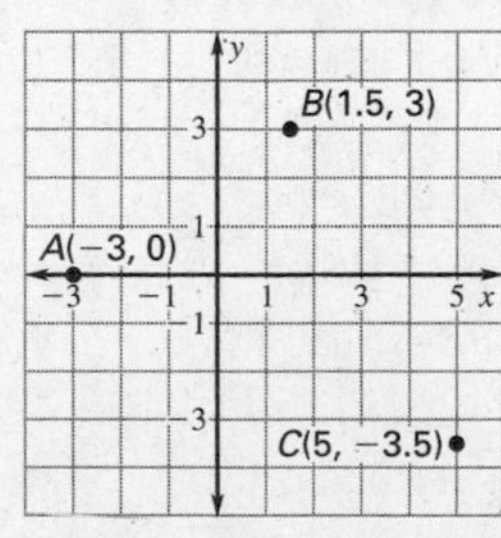

9. x-intercept $= 5$
y-intercept $= 25$

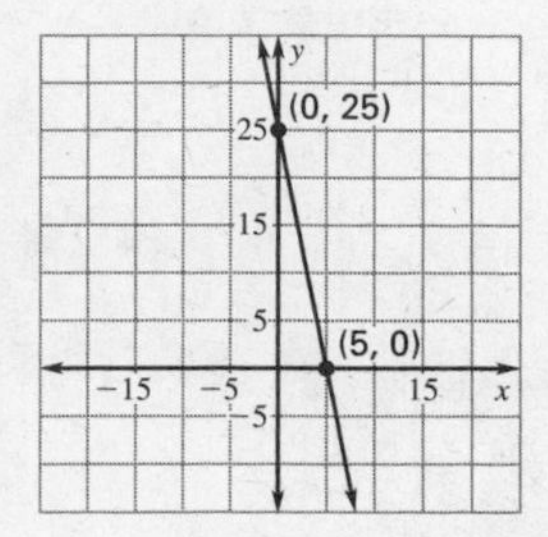

10. x-intercept $= 0.825$
y-intercept $= 3$

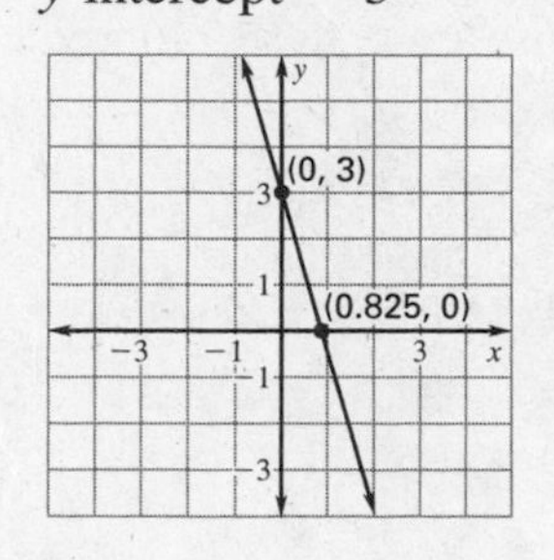

11. x-intercept $= -7$
y-intercept $= -\frac{35}{6}$

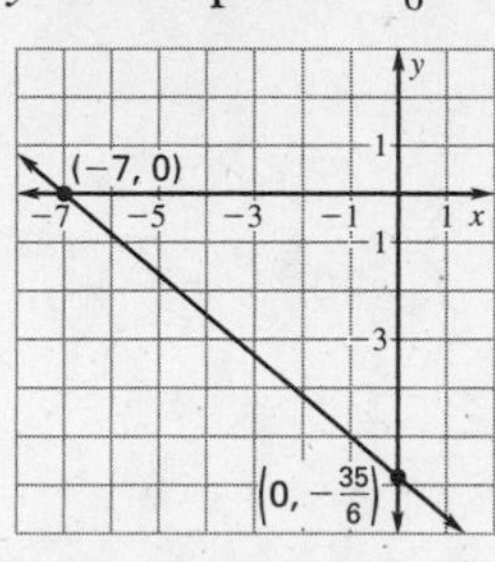

12. x-intercept $= 10$
y-intercept $= \frac{5}{4}$

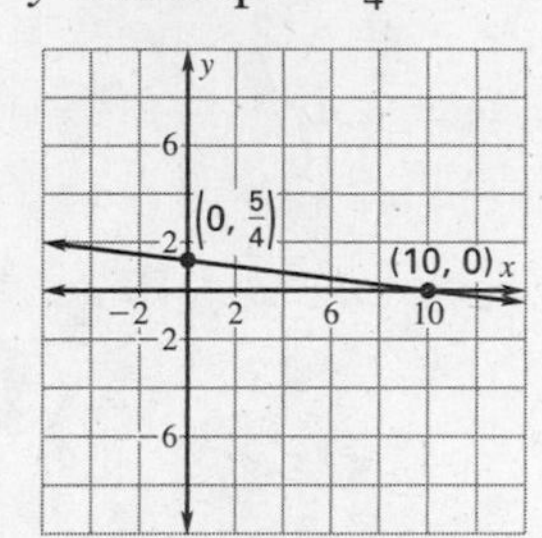

Strategies for Reading Mathematics

1. \$635; \$1035; \$200 per week **2.** \$35
3. $y = 225x + 40$
4. Total earnings = Earnings per extra chore × Number of extra chores + Allowance;
Total earnings = y (dollars),
Earnings per extra chore = 1.50 (dollars),
Number of extra chores = x (chores),
Allowance = 3 (dollars); $y = 1.5x + 3$

Lesson 5.1

Warm-Up Exercises

1. $\frac{1}{3}$ **2.** 3 **3.** $-10, -7, -1$
4. 173, 205, 269

Daily Homework Quiz

1. Yes; no vertical line passes through two points of the graph. **2.** 5, 2, -7 **3.** -1

Lesson Opener

Allow 10 minutes.

1. a. Check graphs. **b.** 1 **c.** 2 **d.** The y-intercept is the constant term in the equations and the slope is the coefficient of x. **2. a.** Check graphs **b.** -3 **c.** 1 **d.** The y-intercept is the constant term in the equation and the slope is the coefficient of x. **3. a.** Check graphs. **b.** 2 **c.** $-\frac{1}{2}$ **d.** The y-intercept is the constant term in the equation and the slope is the coefficient of x.
4. slope $= 5$, y-intercept $= -3$

Lesson 5.1 *continued*

Practice A

1. 2; 5 **2.** −4; 1 **3.** 1; −5 **4.** $\frac{1}{2}$; 0 **5.** 2; 3 **6.** 2; $-\frac{3}{2}$ **7.** $y = x$ **8.** $y = -2x + 4$ **9.** $y = -3x - 5$ **10.** $y = 6x - 1$ **11.** $y = 9$ **12.** $y = -6x - 2$ **13.** $y = 2x - 8$ **14.** $y = -4x + 11$ **15.** $y = 5x + 5$ **16.** $y = -5x - 4$ **17.** $y = -\frac{3}{5}x + 3$ **18.** $y = \frac{8}{9}x - \frac{1}{2}$ **19.** $y = x + 1$ **20.** $y = -x$ **21.** $y = 2$ **22.** $y = -x + 4$ **23.** $y = -2x - 4$ **24.** $y = \frac{3}{2}x - 3$ **25.** $y = 2x$ **26.** 62 in., 68 in., 72 in., 75 in.

Practice B

1. $y = 2x + 3$ **2.** $y = 5x$ **3.** $y = 4x - 3$ **4.** $y = -5x + 1$ **5.** $y = -3x - 2$ **6.** $y = -5$ **7.** $y = \frac{1}{2}x - 8$ **8.** $y = -\frac{3}{4}x + 9$ **9.** $y = -\frac{1}{5}x + 3$ **10.** $y = \frac{4}{5}x - 7$ **11.** $y = \frac{1}{3}x + \frac{2}{3}$ **12.** $y = -\frac{4}{3}x + \frac{7}{8}$ **13.** $y = x + 2$ **14.** $y = -x + 3$ **15.** $y = 2x + 4$ **16.** $y = x - 4$ **17.** $y = \frac{1}{2}x + 1$ **18.** $y = -\frac{3}{2}x + 3$ **19.** $y = 0.005x$ **20.** 0.75, 7.5, 60, 1000 **21.** $y = 0.15x + 29$ **22.** \$32.75, \$36.50, \$44, \$59

Practice C

1. $y = -8x + 5$ **2.** $y = 13x$ **3.** $y = x - 4$ **4.** $y = 7$ **5.** $y = -\frac{1}{4}x + 3$ **6.** $y = \frac{7}{8}x + 8$ **7.** $y = \frac{5}{9}x - 2$ **8.** $y = -\frac{3}{10}x + 10$ **9.** $y = -\frac{7}{11}x + \frac{1}{6}$ **10.** $y = \frac{9}{16}x - \frac{4}{3}$ **11.** $y = 2x - 4$ **12.** $y = -\frac{1}{4}x + 1$ **13.** $y = -\frac{5}{2}x + 5$ **14.** $y = \frac{7}{3}x - 7$ **15.** $y = -3x + 4$ **16.** $y = 4x - 2$ **17.** $y = 12x + 50$ **18.** \$50, \$62, \$74, \$86, \$98 **19.** $y = 0.13x + 31$ **20.** \$34.25, \$37.50, \$44, \$57 **21.** The graph would be steeper.

Reteaching with Practice

$2x + 5$ **2.** $y = x - 4$ **3.** $y = 2$

5. $C = 50 + 0.30n$

Miles (n)	50	100	200	300
Total charge (C)	65	80	110	140

6. a. $E = 1400 + 30t$ **b.** 1580

Interdisciplinary Application

1. $y = 7x + 2500$; $y = 18x$

2.

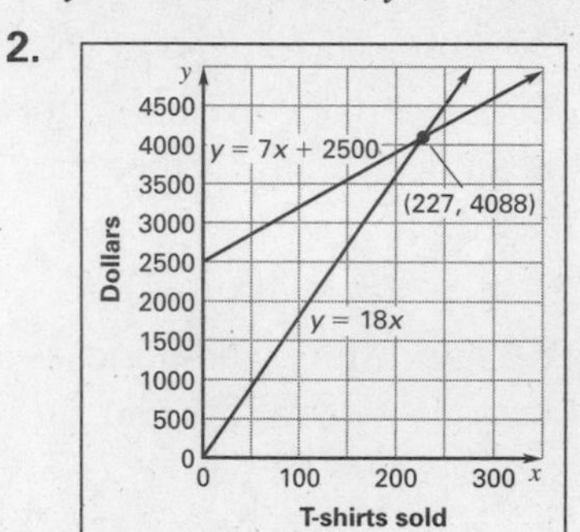

3. 227 T-shirts

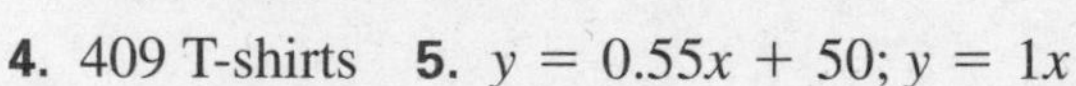

4. 409 T-shirts **5.** $y = 0.55x + 50$; $y = 1x$

6.

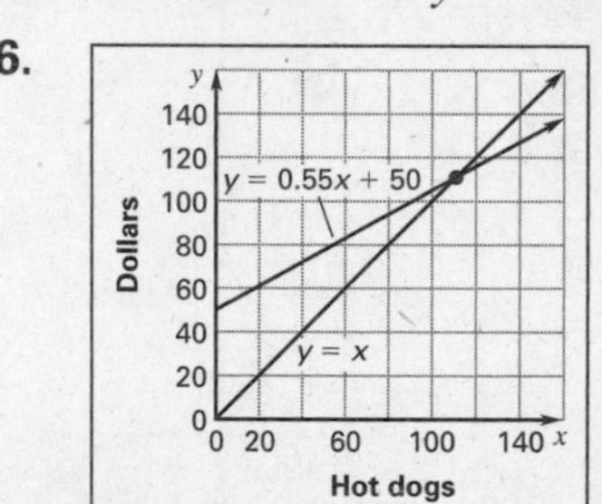

7. 111 hotdogs

Challenge: Skills and Applications

1. 2.45 million; the population of the United States increased an average of 2.45 million people per year between 1950 and 1990. **2.** $y = 2.45t + 151$ **3.** 200 million; 3 million off; *Sample answer*: Yes, it is a fairly close approximation since the error is only about 15%. **4.** 298 million **5.** $y = 23.36t + 285.7$ **6.** \$425.86 billion; \$5.24 billion off; *Sample answer*: It's a fairly close approximation since an error of 5.24 billion out of 431.1 billion is only about a 1% error. **7.** \$519.3 billion; *Sample answer*: The prediction is probably somewhat close since the 1996prediction was fairly close.

Lesson 5.2

Warm-Up Exercises

1. $y = 5x - 2$ **2.** $y = -\frac{2}{3}x + 1$ **3.** $y = -\frac{1}{2}$ **4.** 14 **5.** $-\frac{5}{4}$

Lesson 5.2 *continued*

Daily Homework Quiz

1. $y = -\frac{3}{2}x + 2$

2. $y = x - 2$ 3. $C = 2.5h + 15$

Lesson Opener

Allow 15 minutes.

1. a, b.

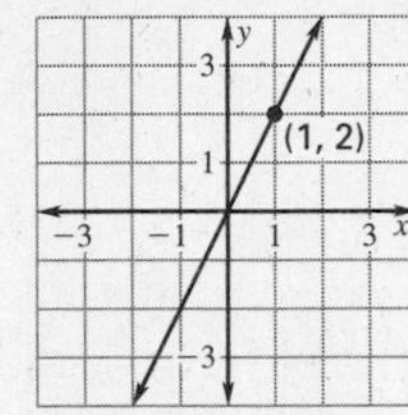

c. 0

d. Yes; the y-intercept and slope are known; $y = 2x$

2. a, b.

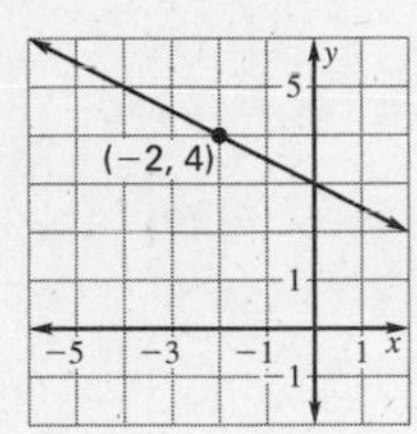

c. 3

d. Yes; the y-intercept and slope are known; $y = -\frac{1}{2}x + 3$

3. a, b.

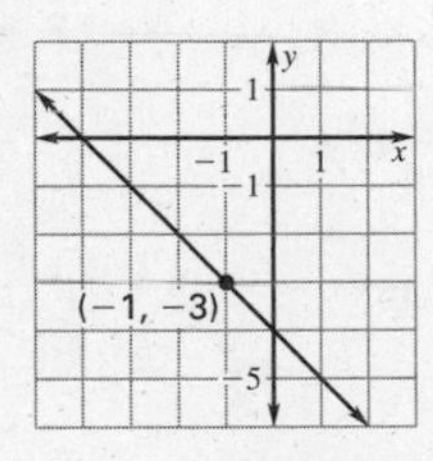

c. -4

d. Yes; the y-intercept and slope are known; $y = -x - 4$

4. a, b.

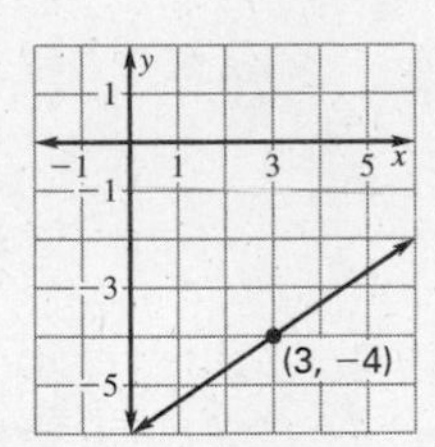

c. -6

d. Yes; the y-intercept and slope are known; $y = \frac{2}{3}x - 6$

5. If the y-intercept and slope are known, the equation of the line can be written.

Practice A

1. $y = x + 2$ 2. $y = 4x + 12$
3. $y = -5x + 7$ 4. $y = 3x - 2$
5. $y = 8x + 33$ 6. $y = -x + 9$
7. $y = -3x + 14$ 8. $y = 10x + 25$
9. $y = \frac{1}{5}x - 11$ 10. $y = 2x + 1$
11. $y = -3x + 2$ 12. $y = -2x - 3$
13. $y = \frac{1}{2}x - 5$ 14. $y = \frac{2}{3}x$ 15. $y = 2$
16. $y = 3x + 2$ 17. $y = x + 1$
18. $y = -3x + 1$ 19. $y = 20t + 310$
20. $y = 15t + 55$

Practice B

1. $y = -x + 8$ 2. $y = 4x + 14$
3. $y = -3x + 19$ 4. $y = 8$ 5. $y = 2x + 6$
6. $y = -7x$ 7. $y = -\frac{5}{3}x - 2$ 8. $y = \frac{3}{4}x + \frac{11}{4}$
9. $y = -\frac{5}{7}x + \frac{1}{7}$ 10. $y = \frac{1}{2}x + \frac{11}{2}$
11. $y = -\frac{1}{3}x - \frac{4}{3}$ 12. $y = \frac{2}{3}x + \frac{8}{3}$ 13. $y = -x$
14. $y = 2$ 15. $x = 5$ 16. $y = 5x - 13$
17. $y = -2x + 10$ 18. $y = \frac{2}{3}x + \frac{1}{3}$
19. $y = 27t + 186$ 20. \$429 21. $y = 21t + 4$
22. 151 stamps 23. $w = 0.75n + 8$
24. \$8 per hour

Practice C

1. $y = 4x - 28$ 2. $y = -\frac{1}{3}x - 5$
3. $y = x + 4$ 4. $y = 5x + 9$ 5. $y = \frac{1}{3}x - 7$
6. $y = -\frac{1}{4}$ 7. $y = -14x - 8$
8. $y = -\frac{3}{5}x + 45$ 9. $y = -\frac{1}{2}x - 7$
10. $y = \frac{2}{5}x + \frac{8}{5}$ 11. $y = -\frac{3}{4}x$ 12. $y = 5$
13. $y = \frac{1}{3}x + \frac{8}{3}$ 14. $y = -\frac{2}{5}x + \frac{4}{5}$ 15. $x = -3$
16. $y = 3x + 3$ 17. $y = \frac{1}{3}x - 2$
18. $y = \frac{1}{2}x + \frac{5}{2}$ 19. $y = 35t + 175$ 20. \$490
21. $y = 35t + 65$ 22. 310 stamps
23. $s = 1500n + 16{,}500$ 24. \$25,500

Reteaching with Practice

1. $y = -2x - 4$ 2. $y = 4x + 10$
3. $y = -x + 9$ 4. $y = 4x - 5$
5. $y = x + 3$ 6. $y = -2x + 1$ 7. \$12.75

Real-Life Application

1. Boys soccer $y = 12x + 34$; Girls soccer $y = 8x + 48$; Football $y = -3x + 119$

2.

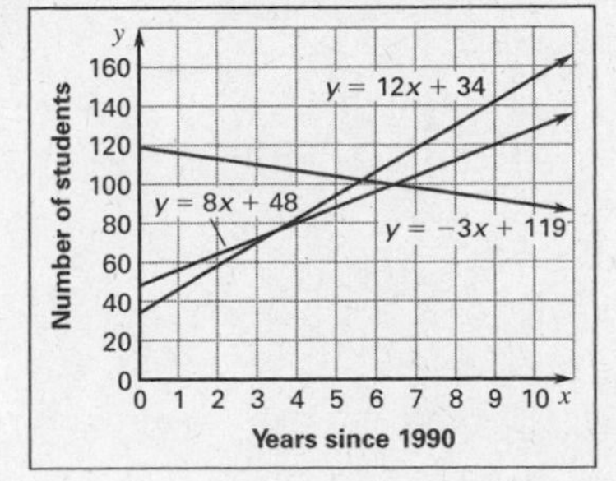

Lesson 5.2 *continued*

3. Boys soccer, 166 people; Girls soccer, 136 people; Football, 86 people

4. Boys soccer, \$106,958.76; Girls soccer, \$87,628.87; Football, \$55,412.37

Challenge: Skills and Applications

1. 9 **2.** 11 **3.** $\frac{17}{2}$ **4.** $-\frac{14}{3}$ **5.** $3k + 4$

6. $k + \frac{h}{r}$ **7.** $y = \left(m + \frac{1}{2}\right)x + b$

8. $y = \frac{2}{3}x + \frac{19}{3}$ **9.** $y = px + q - p^2$

10. $y = 2100x + 33{,}600$ **11.** \$50,400

12. \$65,100

Lesson 5.3

Warm-Up Exercises

1. $y = \frac{1}{2}x + 6$ **2.** $y = -2x + 9$

3. $y = -3x + 7$ **4.** $y = \frac{1}{4}x + \frac{1}{2}$

Daily Homework Quiz

1. a. $y = -x + 4$ **b.** $y = 4x - 8$

2. $y = 2x - 9$

3. a. $P = -150t + 76{,}500$ **b.** 74,550

Lesson Opener

Allow 10 minutes.

1. (5, 30), (3, 18) **2.**

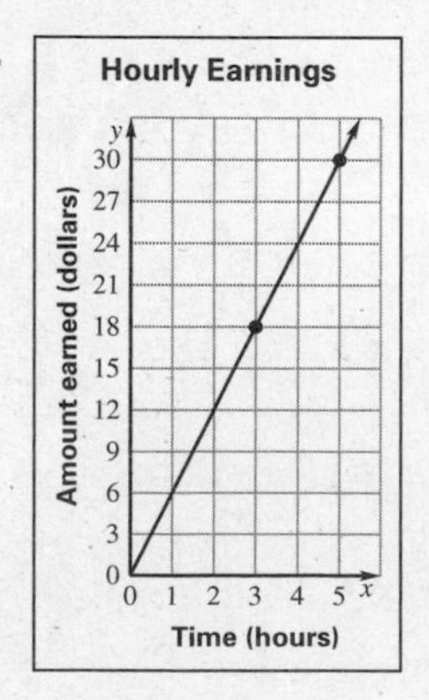

3. 6 **4.** 0 **5.** (2, 14), (5, 20)

6.

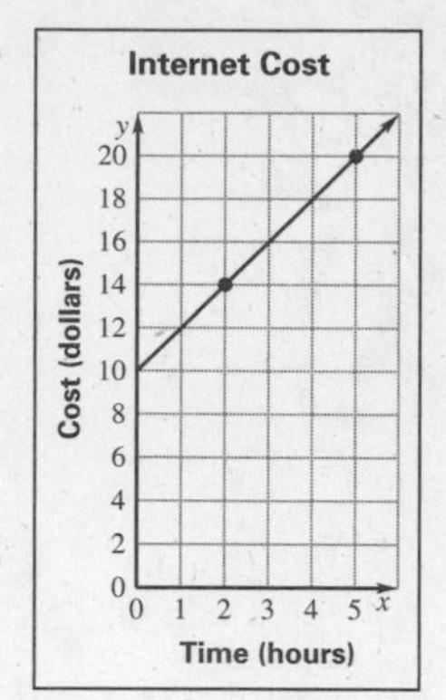

7. 2 **8.** 10

9. Yes; you know the slope and the y-intercept so you can use the slope-intercept form of the line to find the equation.

Graphing Calculator Activity

1. a. -3 **b.** $\frac{1}{2}$ **c.** $-\frac{2}{3}$ **2.** *Sample answer:* $y = \frac{1}{2}x + 5$, $y = -2x$, $y = -2x + 5$

Practice A

1. $y = x + 3$ **2.** $y = x$ **3.** $y = -2x + 1$

4. $y = \frac{1}{4}x - 2$ **5.** $y = 2x - 1$

6. $y = -3x - 4$ **7.** $y = -2x$ **8.** $y = x + 4$

9. $y = -3x - 6$ **10.** $y = x + 2$

11. $y = \frac{1}{3}x - \frac{5}{3}$ **12.** $y = x$ **13.** $y = -4x - 3$

14. $y = 3x + 16$ **15.** $y = 2x - 1$ **16.** -1

17. $\frac{1}{3}$ **18.** -2 **19.** $y = 4x$ **20.** $y = \frac{18}{7}t + 51$

21. $y = -\frac{1}{3}x + \frac{8}{3}$ **22.** $y = \frac{11}{15}x + 50$; $\frac{11}{15}$

Practice B

1. $y = x - 3$ **2.** $y = -3x - 5$

3. $y = 4x - 3$ **4.** $y = -\frac{1}{3}x + 2$

5. $y = 2x - 1$ **6.** $y = -\frac{3}{2}x + 3$

7. $y = 5x + 8$ **8.** $y = -6x - 33$

9. $y = \frac{3}{4}x - \frac{1}{4}$ **10.** $y = -3x + 14$

11. $y = 2x + 4$ **12.** $y = 5x + 31$

13. $y = x - \frac{3}{2}$ **14.** $y = -0.5x - 0.64$

15. $y = -\frac{17}{5}x + \frac{13}{10}$ **16.** $-\frac{1}{3}$ **17.** $\frac{3}{2}$ **18.** $\frac{1}{2}$

19. Slope $\overline{ZW} = -\frac{2}{5}$, slope $\overline{ZY} = \frac{5}{2}$; $\overline{ZY}$ and $\overline{ZW}$ are perpendicular since $-\frac{2}{5}$ is the negative reciprocal of $\frac{5}{2}$. Slope $\overline{YX} = -\frac{2}{5}$, slope $\overline{ZY} = \frac{5}{2}$; $\overline{YX}$ and $\overline{ZY}$ are perpendicular since $-\frac{2}{5}$ is the negative reciprocal of $\frac{5}{2}$. **20.** line through

Lesson 5.3 *continued*

$\overline{ZW}$: $y = -\frac{2}{5}x + \frac{16}{5}$; line through $\overline{ZY}$: $y = \frac{5}{2}x + 9$; line through $\overline{YX}$: $y = -\frac{2}{5}x - \frac{13}{5}$
21. line through $\overline{ZW}$: $y = -\frac{2}{5}x + \frac{16}{5}$; line through $\overline{YX}$: $y = -\frac{2}{5}x - \frac{13}{5}$; $\overline{ZW}$ and $\overline{YX}$ are parallel since their slopes are equal. **22.** $y = 60t - 20$

Practice C

1. $y = \frac{1}{2}x - 1$ **2.** $y = 4x - 3$
3. $y = -x - 5$ **4.** $y = -\frac{5}{2}x + 12$
5. $y = -\frac{3}{2}x - \frac{21}{2}$ **6.** $y = \frac{4}{7}x - \frac{13}{7}$
7. $y = 6x + 4$ **8.** $y = 5x + 8$ **9.** $y = \frac{1}{3}x - 4$
10. $y = -6x + \frac{23}{2}$ **11.** $y = -\frac{3}{4}x - 2$
12. $y = \frac{1}{4}x + 7$ **13.** $y = -4.7$ **14.** $y = 3x$
15. $y = -0.5x + 7$ **16.** $\frac{1}{7}$ **17.** $\frac{3}{2}$ **18.** $\frac{5}{3}$
19. Slope $\overline{ZW} = -\frac{2}{5}$, slope $\overline{ZY} = \frac{5}{2}$; $\overline{ZY}$ and $\overline{ZW}$ are perpendicular since $-\frac{2}{5}$ is the negative reciprocal of $\frac{5}{2}$. Slope $\overline{YX} = -\frac{2}{5}$, slope $\overline{ZY} = \frac{5}{2}$; $\overline{YX}$ and $\overline{ZY}$ are perpendicular since $-\frac{2}{5}$ is the negative reciprocal of $\frac{5}{2}$. **20.** line through $\overline{ZW}$: $y = -\frac{2}{5}x + \frac{16}{5}$; line through $\overline{ZY}$: $y = \frac{5}{2}x + 9$; line through $\overline{YX}$: $y = -\frac{2}{5}x - \frac{13}{5}$
21. line through $\overline{ZW}$: $y = -\frac{2}{5}x + \frac{16}{5}$; line through $\overline{YX}$: $y = -\frac{2}{5}x - \frac{13}{5}$; $\overline{ZW}$ and $\overline{YX}$ are parallel since their slopes are equal.
22. $y = -12.97t + 132.97$ **23.** 16.24 million

Reteaching with Practice

1. $y = x + 5$ **2.** $y = -8x + 7$
3. $y = 3x + 3$ **4.** $y = -\frac{1}{2}x + 5$
5. $y = 3x - 14$ **6.** $y = \frac{1}{4}x + 2$

Interdisciplinary Application

1. $y = 393x + 14324$
2.

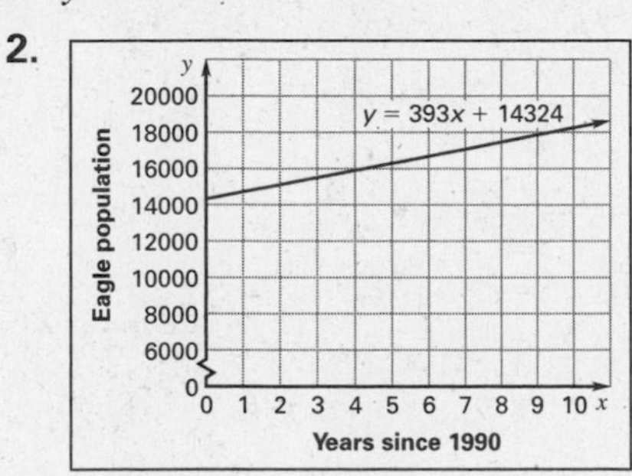

3. 18,254 **4.** *Sample answer*: Habitat loss, less people counting. **5.** $y = 518.25x + 13698$
6.

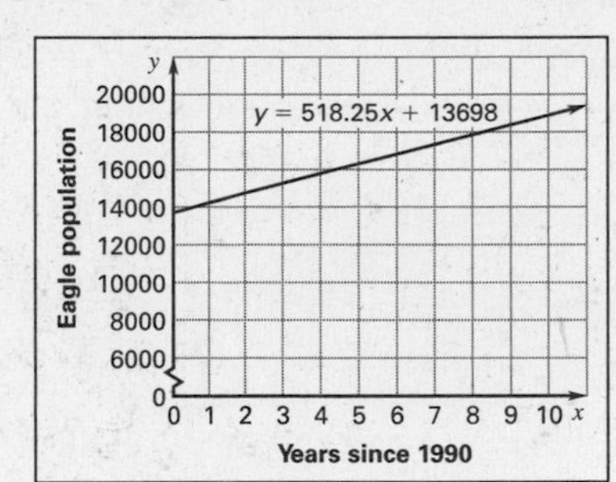

7. 18,880

Challenge: Skills and Applications

1. $y = -\frac{20}{9}x$ **2.** $y = -\frac{1}{2}x + \frac{7}{12}$
3. $y = 3x - 3k - 2$ **4.** $y = 6x - 17$
5. $y = -\frac{1}{6}x + \frac{3}{2}$ **6.** $y = 6x + 20$
7. $y = -\frac{1}{6}x + \frac{23}{3}$ **8.** $y = 0.625x - 2.75$, or $y = \frac{5}{8}x - \frac{11}{4}$ **9.** about 24 days **10.** 9.75 in.
11. -2.75 inches; *Sample answers*: One possibility is that the seed is planted $2\frac{3}{4}$ in. below ground so it starts off with a height of $-2\frac{3}{4}$ in. Another possibility is that the linear growth model in Exercise 8 is not appropriate until the plant has reached a certain stage of growth. Extending the line to the left results in a negative value for $x = 0$, but the values obtained from the model may not reflect the real-life situation for the first few days.

Quiz 1

1. $y = -3x + 7$ **2.** $y = -\frac{3}{5}x - 3$
3. $y = 4x + 3$ **4.** $y = \frac{2}{3}x - \frac{2}{3}$
5. $y = 2x + 10$ **6.** $y = \frac{1}{4}x + 5$ **7.** $y = \frac{1}{5}x + 6$

Lesson 5.4

Warm-Up Exercises

1. $y = -8x + 72$ **2.** $y = 18x + 160$
3. $y = -\frac{1}{2}x - 21$

Daily Homework Quiz

1. a. $y = -\frac{3}{5}x - \frac{4}{5}$ **b.** $y = \frac{1}{2}x + 2$
2. $y = -3x + 5$
3. The slope of $\overline{AC}$ is $\frac{1}{2}$; the slope of $\overline{BC}$ is -2. Therefore, $\overline{AC}$ is perpendicular to $\overline{BC}$.

Lesson 5.4 *continued*

Lesson Opener

Allow 10 minutes.

1.

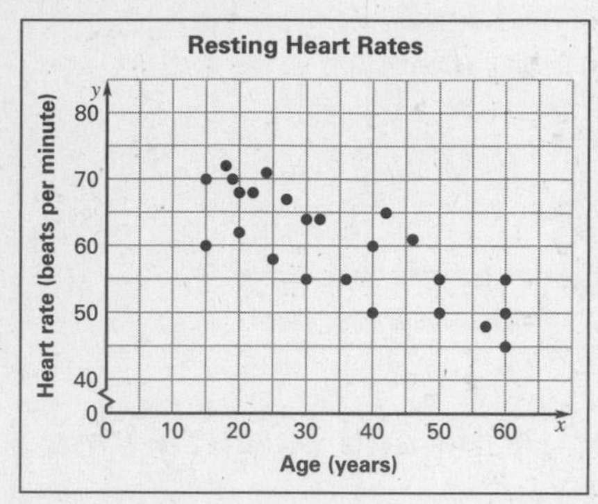

2. b; This equation has a slope and a y-intercept that best fits the data.

Practice A

1. yes **2.** no **3.** yes **4–12.** Sample answers are given. **4.** $y = x$ **5.** $y = x - 4$

6. $y = -x + 2$

7.

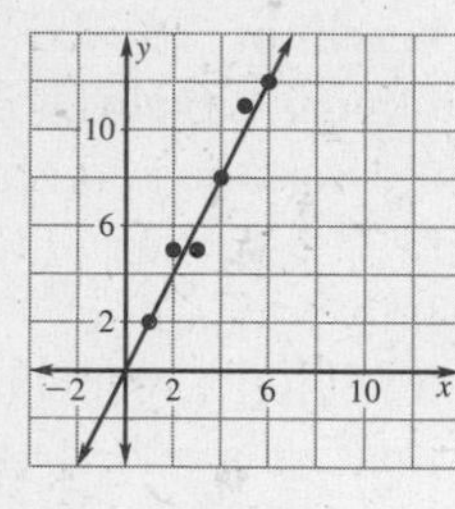

$y = 2x$

8.

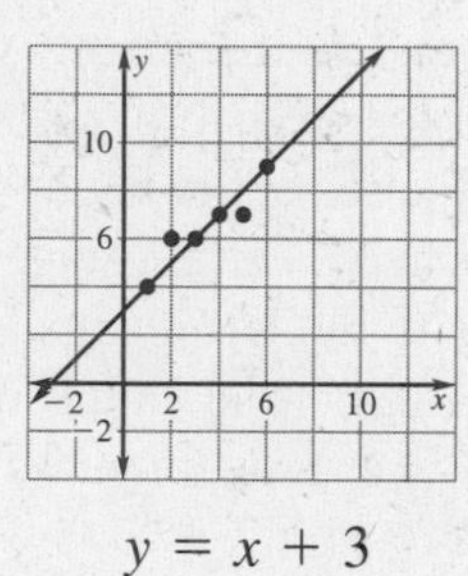

$y = x + 3$

9.

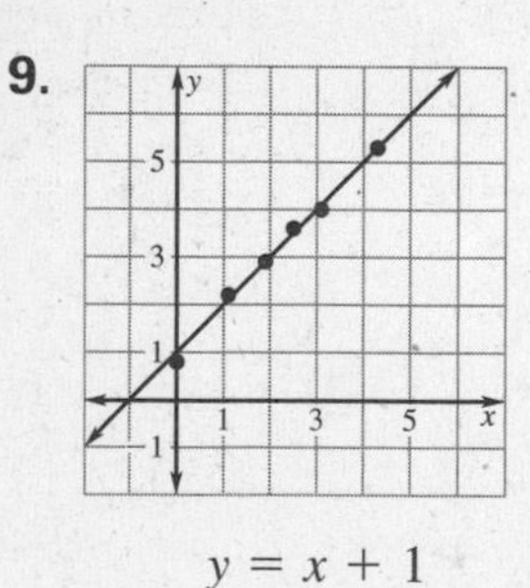

$y = x + 1$

10.

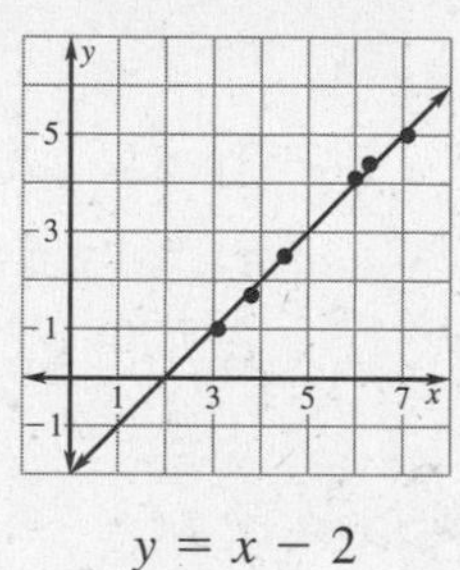

$y = x - 2$

11. $y = 180 - 2x$; 160 lb

12. $y = \frac{4}{5}x + \frac{1}{5}$, 5.8 gal

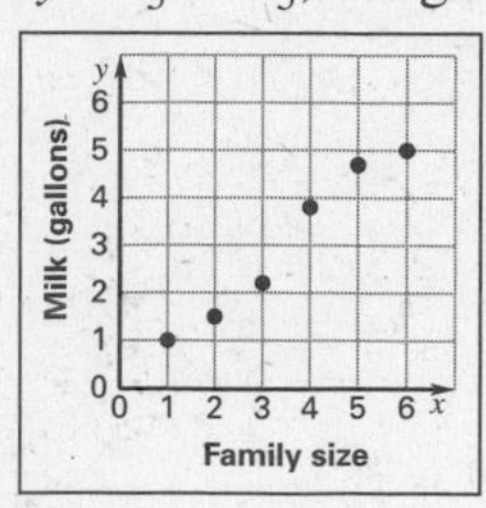

Practice B

1–3. Answers will vary. **4–14.** Sample answers are given. **4.** $y = 2x + 2$ **5.** $y = -2x$

6. $y = -\frac{1}{2}x - 3$

7.

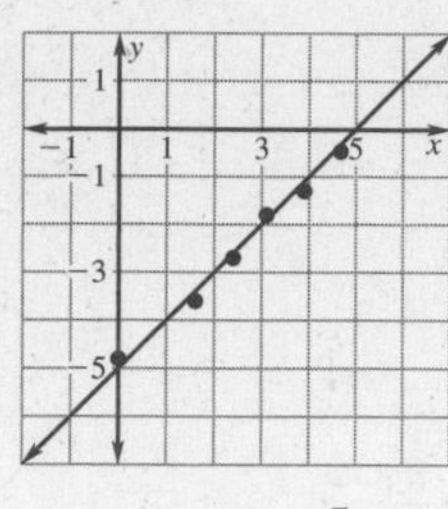

$y = x - 5$

8.

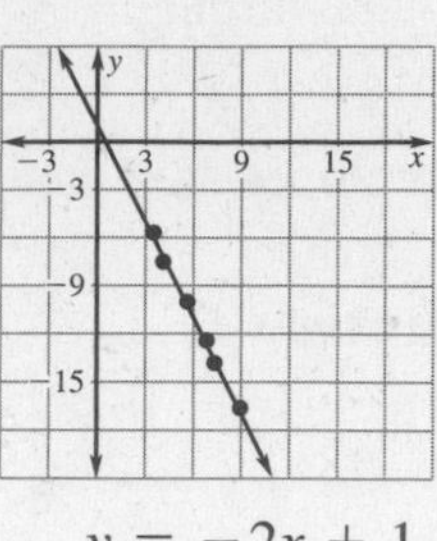

$y = -2x + 1$

9.

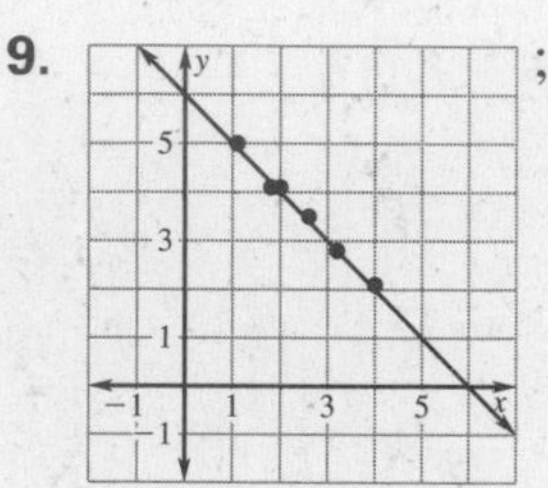

; $y = -x + 6$

10.

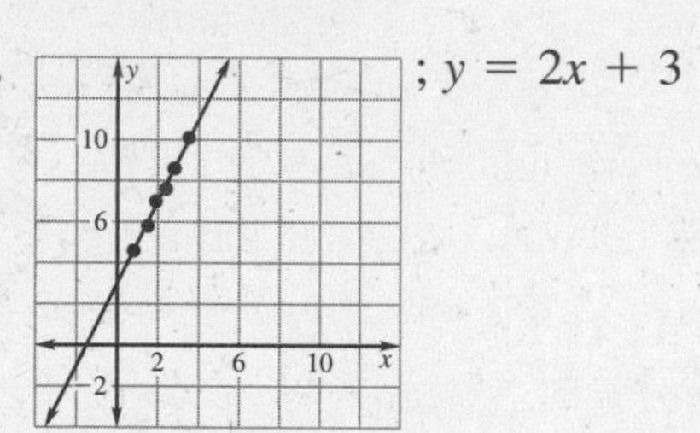

; $y = 2x + 3$

11. $y = 180 - 2x$; 160 lb **12.** no; after one year the person would lose 104 pounds and would weigh less than 80 pounds. **13.** $y = \frac{4}{5}x + \frac{1}{5}$; 5.8 gal **14.** yes; it seems reasonable that it would continue.

Practice C

1–13. Sample answers are given.

1. $y = 3x - 2$ **2.** $y = \frac{1}{3}x + 2$

3. $y = -\frac{3}{5}x - \frac{4}{5}$

4.

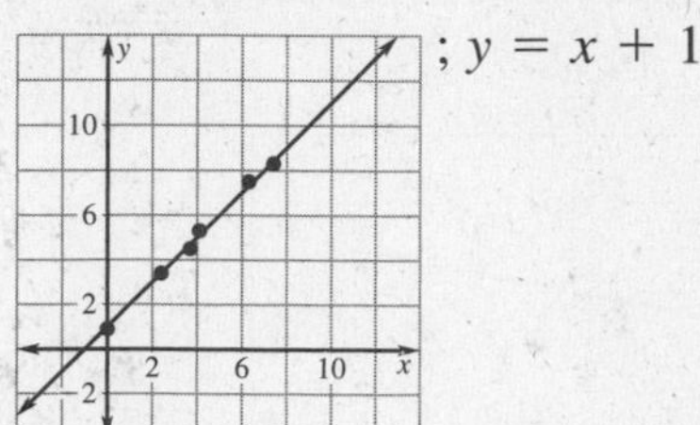

; $y = x + 1$

5. 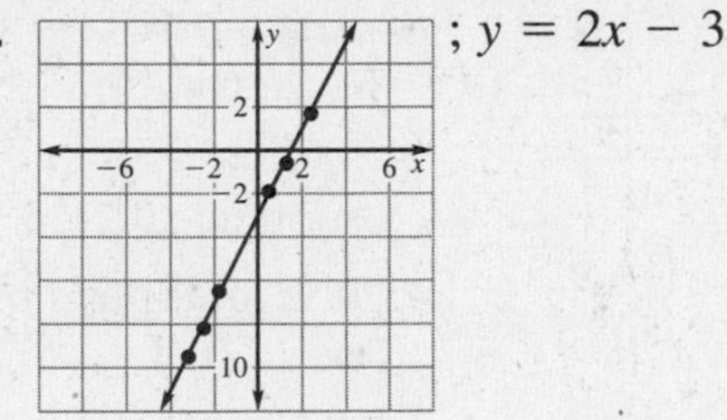

; $y = 2x - 3$

6. 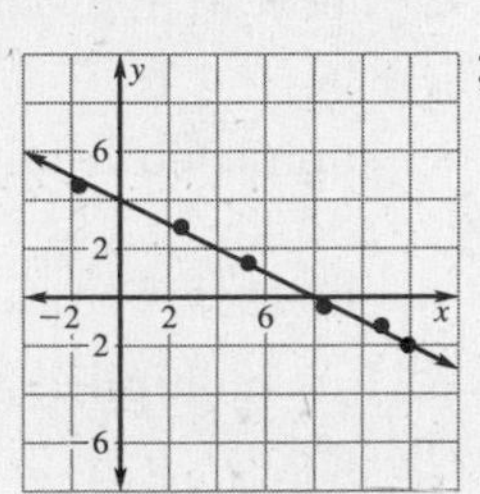

; $y = -\frac{1}{2}x + 4$

7. 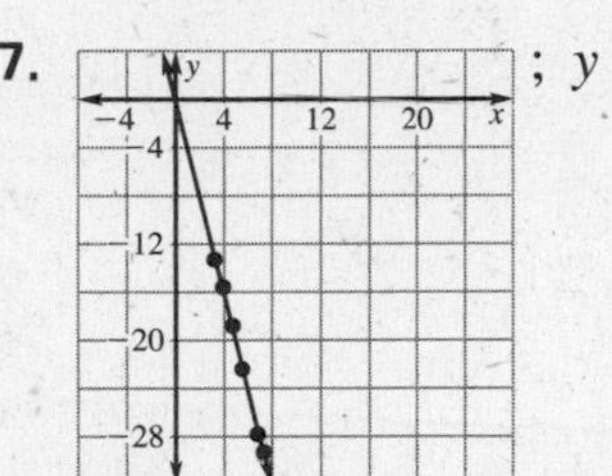

; $y = -4x - \frac{1}{4}$

8. $y = 180 - 2x$; 160 lb

9. The slope is weight loss per week.

10. No; after one year the person would lose 104 pounds and would weigh less than 80 pounds.

11. $y = \frac{4}{5}x + \frac{1}{5}$, 5.8 gal

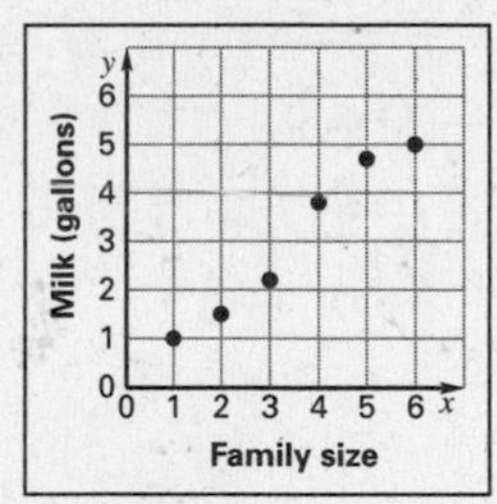

12. The slope is milk consumption per person.

13. Yes; it seems reasonable that it would continue.

Reteaching with Practice

1.

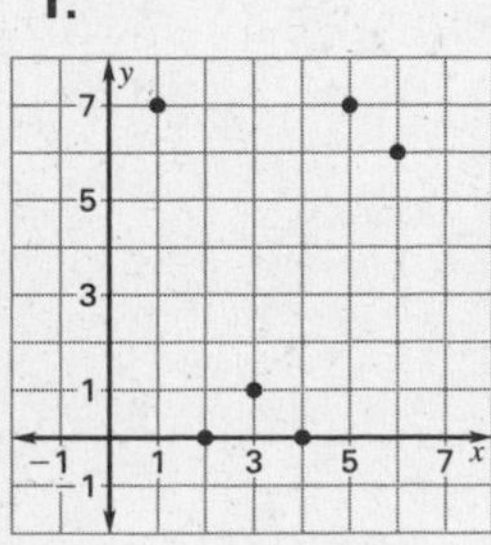

no best-fitting line

2.

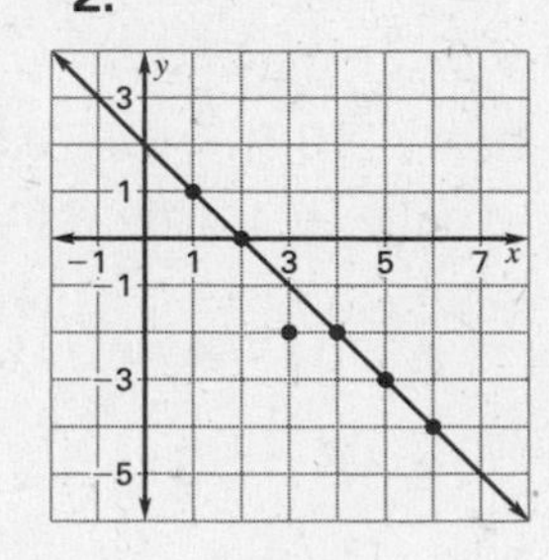

$y = -x + 2$

3. negative correlation **4.** no correlation

Cooperative Learning Activity

1. Height, forearm length, and age should all have positive correlation when compared to each other. None of the data should show negative correlation. Comparing test scores with any of the other data should reflect no correlation.

2. *Sample answers:* Comparing height and weight is an example of positive correlation. Comparing the outside temperature and a heating bill is an example of negative correlation. Comparing a student's shoe size with the number of pets in his or her household is an example of no correlation.

Real-Life Application

1.

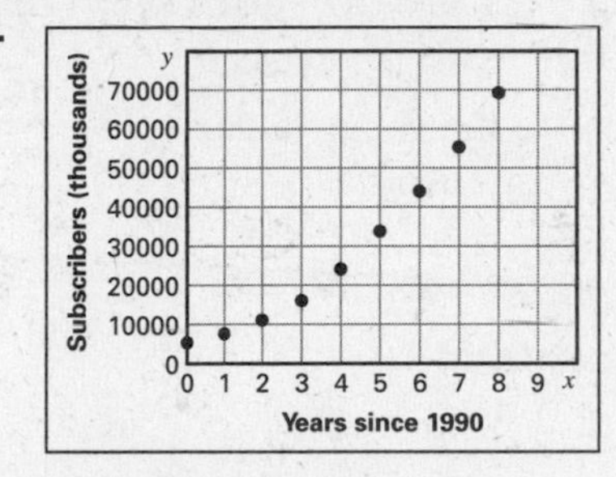

2. $y = 7300x + 257$ **3.** 73,257

4.

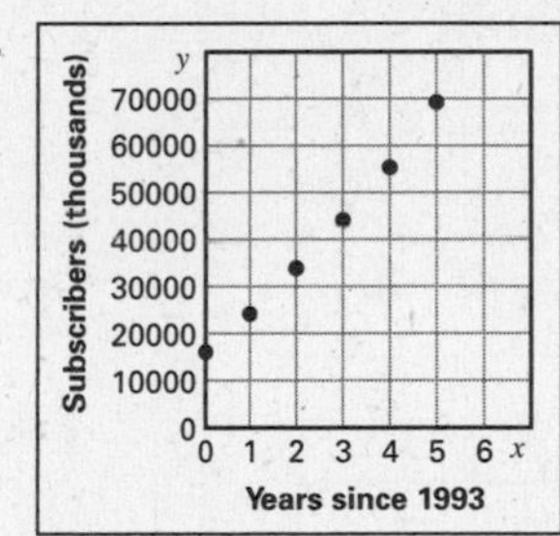

5. $y = 8850x + 16{,}000$ **6.** 77,950

Challenge: Skills and Applications

1. positive correlation; salaries tend to increase as years of experience increase. **2.** no correlation; shoe size is not related to test scores.

3. negative correlation; the longer a person is on the plan, the less he or she should weigh.

4. positive correlation; more time spent training should increase the distance a person can run.

Lesson 5.4 *continued*

5–7. Sample equations given were determined using a graphing calculator and rounding to the nearest hundredth. Sample estimate was found using the given sample equation.

5. *Sample answer*: $y = 0.29x + 50.82$

6. *Sample answer*: 79.82 years

7. *Sample answer*: $y = 1.31x - 17.46$ **8.** 21

9. 1930 and 1940: 0.29; 1930 and 1950: 0.79; 1930 and 1960: 0.83; 1930 and 1970: 1.04; 1930 and 1980: 1.22; 1930 and 1990: 1.24; 1940 and 1950: 1.29; 1940 and 1960: 1.10; 1940 and 1970: 1.29; 1940 and 1980: 1.46; 1940 and 1990: 1.43; 1950 and 1960: 0.9; 1950 and 1970: 1.29; 1950 and 1980: 1.51; 1950 and 1990: 1.47; 1960 and 1970: 1.67; 1960 and 1980: 1.82; 1960 and 1990: 1.66; 1970 and 1980: 1.97; 1970 and 1990: 1.65; 1980 and 1990: 1.33 **10.** about 1.30

11. They are close to the same value.

Lesson 5.5

Warm-Up Exercises

1. $-\frac{4}{7}$ **2.** $\frac{3}{7}$ **3.** $y = \frac{3}{5}x + 27$

4. $y = -2x + 5$

Daily Homework Quiz

1. a. and **b.**

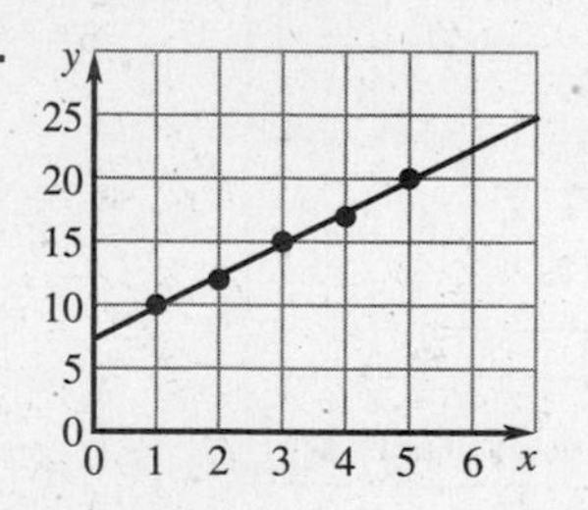

2. *Sample answer:* $y = 2.5x + 7.3$

3. positive correlation

Lesson Opener

Allow 15 minutes.

1. a, b.

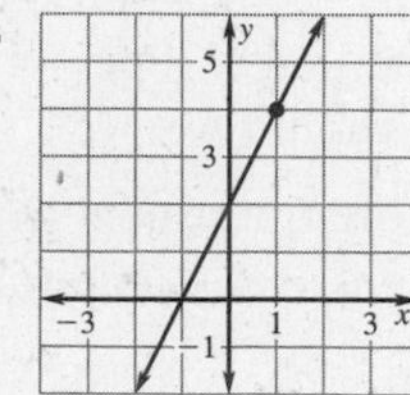

c. They are the same.

d. The 1 is subtracted from x and the 4 is subtracted from y.

e. It is multiplied by $(x - 1)$.

2. a, b.

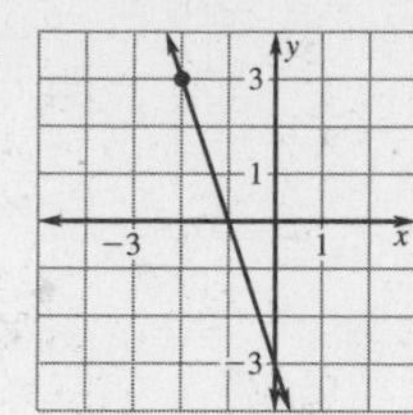

c. They are the same.

d. The -2 is subtracted from x and the 3 is subtracted from y.

e. It is multiplied by $(x + 2)$.

3. a, b.

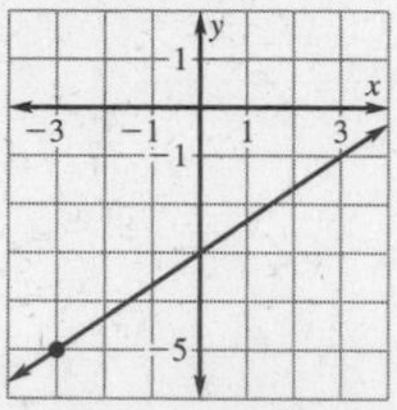

c. They are the same.

d. The -3 is subtracted from x and the -5 is subtracted from y.

e. It is multiplied by $(x + 3)$.

Practice A

1. 6 **2.** 1 **3.** 2 **4.** 2 **5.** $-\frac{1}{2}$ **6.** 0

7. $y - 5 = 2(x - 1)$ or $y + 1 = 2(x + 2)$

8. $y - 2 = -3(x + 1)$ or $y + 4 = -3(x - 1)$

9. $y - 2 = \frac{1}{3}(x - 3)$ or $y = \frac{1}{3}(x + 3)$

10. $y - 5 = 3(x - 2)$ **11.** $y - 4 = 2(x - 1)$

12. $y = \frac{1}{2}(x + 2)$ **13.** $y - 7 = x + 3$

14. $y - 8 = -4(x + 5)$ **15.** $y + 4 = 9x$

16. $y - 1 = 0$ **17.** $y + 4 = -2(x + 3)$

18. $y + 10 = 5(x - 6)$

19. $y = 5x$ or $y - 5 = 5(x - 1)$

20. $y - 3 = 2(x - 2)$ or $y - 7 = 2(x - 4)$

21. $y - 6 = 3(x - 9)$ or $y + 6 = 3(x - 5)$

22. $y + 7 = -(x - 8)$ or $y + 8 = -(x - 9)$

23. $y + 2 = 7(x - 1)$ or $y - 5 = 7(x - 2)$

24. $y + 7 = \frac{1}{2}(x - 2)$ or $y + 10 = \frac{1}{2}(x + 4)$

25. $y = 2x + 1$ **26.** \$13 **27.** $d = \frac{13}{14}t$

28. 6:42 P.M.

Lesson 5.5 *continued*

Practice B

1. $y + 2 = 2(x + 1)$ or $y - 2 = 2(x - 1)$
2. $y - 1 = \frac{2}{5}(x - 2)$ or $y + 1 = \frac{2}{5}(x + 3)$
3. $y + 1 = -\frac{1}{2}(x + 4)$ or $y + 4 = -\frac{1}{2}(x - 2)$
4. $y - 24 = -2(x + 3)$ **5.** $y + 2 = -5(x + 4)$
6. $y + 3 = \frac{2}{3}x$ **7.** $y + 5 = -4(x - 6)$
8. $y - 1 = 0$ **9.** $y + 5 = 6(x + 3)$
10. $y - 5 = -2x$ or $y - 3 = -2(x - 1)$
11. $y - 4 = -\frac{1}{2}(x - 2)$ or $y - 2 = \frac{1}{2}(x - 6)$
12. $y = x - 3$ or $y + 3 = x$
13. $y + 2 = \frac{3}{4}(x - 6)$ or $y - 1 = \frac{3}{4}(x - 10)$
14. $y - 1 = \frac{2}{3}(x - 4)$ or $y + 3 = \frac{2}{3}(x + 2)$
15. $y - 3 = x - 1$ or $y + 3 = x + 5$
16. $y + 7 = -\frac{3}{2}(x + 5)$ or $y + 10 = -\frac{3}{2}(x + 3)$
17. $y - 11 = \frac{9}{7}(x - 6)$ or $y - 2 = \frac{9}{7}(x + 1)$
18. $y - 4 = \frac{12}{5}(x - 2)$ or $y + 8 = \frac{12}{5}(x + 3)$
19. $y = 5x + 6$ **20.** $y = -2x + 1$
21. $y = 3x - 7$ **22.** $y = -3x + 16$
23. $y = \frac{1}{2}x - 12$ **24.** $y = 4x + \frac{7}{3}$
25. $y = 1.5x + 2$ **26.** \$11 **27.** $y = 7x$
28. about 9:46 A.M.

Practice C

1. $y - 1 = -\frac{1}{2}(x - 2)$ or $y - 3 = -\frac{1}{2}(x + 2)$
2. $y - 1 = \frac{3}{2}(x - 2)$ or $y + 5 = \frac{3}{2}(x + 2)$
3. $y + 5 = -\frac{2}{5}(x - 5)$ or $y + 1 = -\frac{2}{5}(x + 5)$
4. $y - 16 = 8(x + 4)$
5. $y + 6 = -\frac{3}{2}(x - 21)$ **6.** $y + 9 = 0$
7. $y + 35 = \frac{7}{9}(x - 18)$
8. $y + 13 = -\frac{1}{4}(x + 12)$
9. $y - 3 = \frac{6}{5}(x + 10)$
10. $y - 6 = -(x + 12)$ or $y + 10 = -(x - 4)$
11. $y + 6 = -9(x - 8)$ or $y - 21 = -9(x - 5)$
12. $y - 7 = -\frac{2}{3}(x - 24)$ or $y - 15 = -\frac{2}{3}(x - 12)$
13. $y + 9 = -\frac{9}{7}(x + 6)$ or $y - 9 = -\frac{9}{7}(x + 20)$
14. $y - 8 = \frac{11}{7}(x + 18)$ or $y - 19 = \frac{11}{7}(x + 11)$
15. $y - 15 = \frac{6}{19}(x + 40)$ or $y - 27 = \frac{6}{19}(x + 2)$
16. $y = 8x + 17$
17. $y = -x + 11$ **18.** $y = -\frac{1}{2}x - 3$
19. $y = 7x - \frac{23}{6}$ **20.** $y = 2x - 11.8$
21. $y = -6x - 9.7$ **22.** $y = 3.25x + 3.5$
23. \$26.25 **24.** $d = \frac{20}{19}t$ **25.** about 4:30 P.M.
26. $y = 2x + 10$

Reteaching with Practice

1. $y = 2x - 3$ **2.** $y = -3x + 3$
3. $y = -4x$ **4.** 8:50 A.M. **5.** $d = -\frac{1}{6}t + 10$

Interdisciplinary Application

1. $y - 45 = 10(x - 3)$
2.

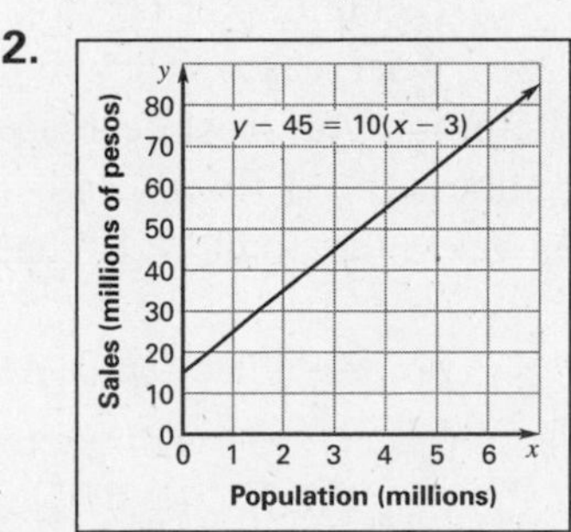

3. 85 million
4. $y - 15 = 7.3(x - 3)$

Lesson 5.5 *continued*

5.

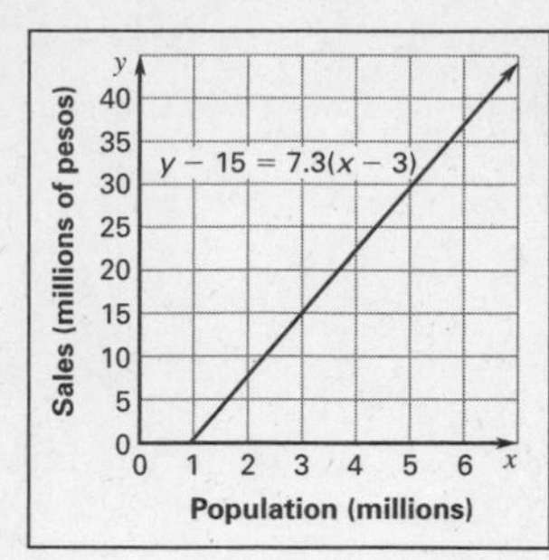

6. 36.9 million pesos

Math and History

1. 5000–5500 years ago; about 50%
2. $y = -(x - 2000)$; $y = -x + 2000$; yes

Challenge: Skills and Applications

1. $y - 11 = 10(x - 2)$ or $y + 4 = 10\left(x - \frac{1}{2}\right)$
2. $y - 3 = \frac{3}{5}(x - 8)$ or $y + 2 = \frac{3}{5}\left(x + \frac{1}{3}\right)$
3. $y - 0.9 = \frac{1}{2}(x + 0.5)$ or $y + 0.5 = \frac{1}{2}(x + 3.3)$
4. $y + 1.4 = -4(x - 3.2)$ or
$y - 1.8 = -4(x - 2.4)$
5. $y - 4 = -2\left(x + \frac{2}{3}\right)$ or $y + \frac{1}{3} = -2\left(x - \frac{3}{2}\right)$
6. $y + 4 = -\frac{5}{6}(x - 5)$ or $y + \frac{1}{4} = -\frac{5}{6}\left(x - \frac{1}{2}\right)$
7. $y - q = -\dfrac{q}{2p}(x - p)$ or $y - 2q = -\dfrac{q}{2p}(x + p)$
8. $y + q = -(x - 2p)$ or $y + q - p = -(x - p)$
9. $y - 3 = -\frac{5}{2}(x - 6)$
10. (4, 8); *Sample explanations*: Method 1: Substitute $2x$ for y in the equation from Exercise 9 and then solve for x; Method 2: Find points on the line by using the slope $-\frac{5}{2}$ to count up or down from (6, 3). As x increases, y decreases so y will not become twice x. Therefore, you must decrease x and increase y. Keep subtracting 1 from x and adding $2\frac{1}{2}$ to y until you get to the point (4, 8) that works.
11. $(12, -12)$ **12.** (2, 13)
13. $y - 5620 = 280(x - 1)$ or
$y - 6040 = 280(x - 2.5)$ **14.** 6740 ft
15. 4780 ft

Quiz 2

1, 2. Check graphs. *Sample answers are given.*
1. $y = \frac{4}{3}x - 2$ **2.** $y = -\frac{2}{3}x + 2$; negative correlation **3.** $y + 7 = \frac{3}{5}(x - 5)$ or $y + 4 = \frac{3}{5}(x - 10)$
4. $y - 4 = \frac{3}{8}(x + 5)$ or $y - 7 = \frac{3}{8}(x - 3)$
5. $y - 1 = \frac{1}{2}(x + 7)$

Lesson 5.6

Warm-Up Exercises

1. $y - 3 = -2(x + 1)$ **2.** $y + 5 = -(x - 3)$ or $y + 4 = -(x - 2)$ **3.** $y = -2x + 9$
4. $y = \frac{1}{2}x + 6$

Daily Homework Quiz

1. $y - 3 = \frac{1}{2}(x - 5)$ or $y = \frac{1}{2}(x + 1)$
2. $y + 2 = -3(x + 4)$ or $y + 17 = -3(x - 1)$
3. $y - 4 = -\frac{5}{2}(x - 6)$

Lesson Opener

Allow 15 minutes.

1, 2. The equations should be matched as follows:
$y = 2x + 1$, $-2x + y = 1$;
$y = 4x - 1$, $4x - y = 1$;
$y = \frac{1}{5}x + 2$, $-x + 5y = 10$;
$y = -x + 3$, $x + y = 3$;
$y = -\frac{1}{2}x - 4$, $x + 2y = -8$;
$y = 5x + 1$, $5x - y = -1$;
$y = x - 6$, $x - y = 6$;
$y = \frac{2}{3}x + 3$, $-2x + 3y = 9$

Graphing Calculator Activity

1. a.

x	−2	−1	0	1	2
y	5	4	3	2	1

The tables are the same; Yes

b.

x	−2	−1	0	1	2
y	$-3.\overline{3}$	−2.5	$-1.\overline{6}$	$-0.8\overline{3}$	0

The tables are not the same; No

c.

x	−2	−1	0	1	2
y	−1.5	−0.625	0.25	1.125	2

The tables are the same; Yes

2. No; Solve the equation for y to put it in slope-intercept form.

Lesson 5.6 *continued*

Practice A

1. $x - y = 9$ **2.** $6x - 4y = -7$ **3.** $x = -7$
4. $2x + 3y = 6$ **5.** $11x + y = -4$ **6.** $y = 1$
7. $4x - y = -3$ **8.** $x - 8y = -2$
9. $x - y = 0$ **10.** $10x - 2y = 1$
11. $x - 4y = -12$ **12.** $2x + 3y = -3$
13. $x - y = -4$ **14.** $3x + y = 11$
15. $-8x + y = 11$ **16.** $4x - y = 31$
17. $2x + y = 16$ **18.** $2x + y = -17$
19. $x - y = 7$ **20.** $5x - y = -3$
21. $2x + y = -4$ **22.** $-6x + y = 1$
23. $3x + y = -9$ **24.** $-4x + y = 10$
25. $2x + y = 30$
26. 30, 20, 14, 10, 0

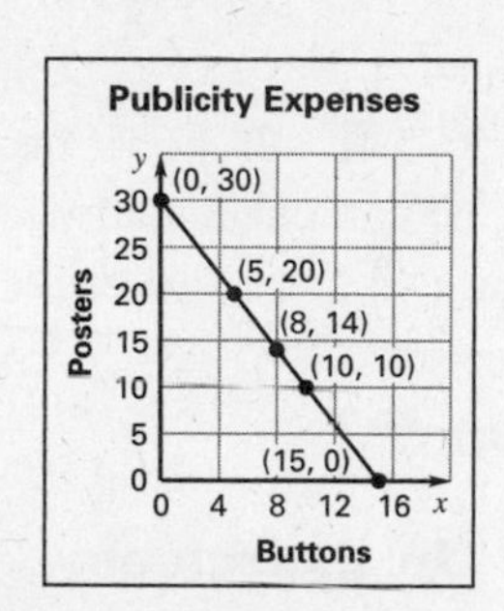

27. $10x + 12y = 240$
28. 20, 15, 10, 5, 0

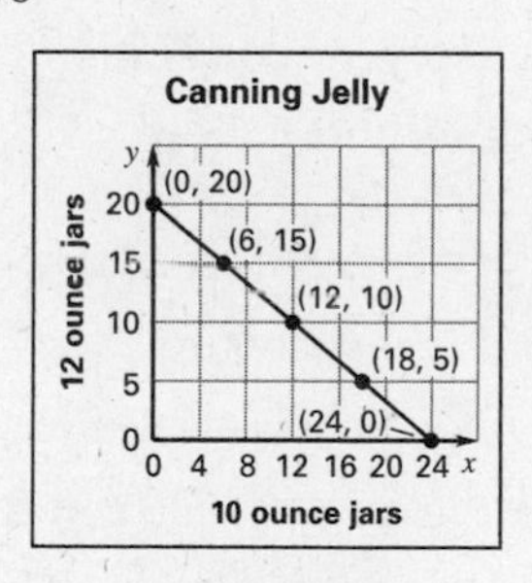

Practice B

1. $2x - y = 8$ **2.** $3x - 4y = 75$
3. $-3x + y = 2$ **4.** $3x + y = 5$
5. $6x - 21y = 18$ **6.** $2x - 3y = 5$ **7.** $x = 4$
8. $y = 4$ **9.** $10x - 3y = 45$
10. $x - 8y = -12$ **11.** $-x + 2y = 8$
12. $-2x + 3y = -5$ **13.** $-2x + y = -5$
14. $4x + y = 9$ **15.** $-3x + y = 6$
16. $6x + y = -8$ **17.** $-x + 3y = -30$
18. $x + 2y = 6$ **19.** $-3x + y = -7$
20. $3x + y = -1$ **21.** $x + 8y = 17$
22. $-5x + y = 15$ **23.** $-x + 2y = -6$
24. $x + 3y = 12$ **25.** $y = -4, x = 3$
26. $y = 1, x = 5$ **27.** $y = -2, x = -3$
28. $y = -4, x = 0$ **29.** $2x + 3y = 60$
30. $y = -\frac{2}{3}x + 20$
31.

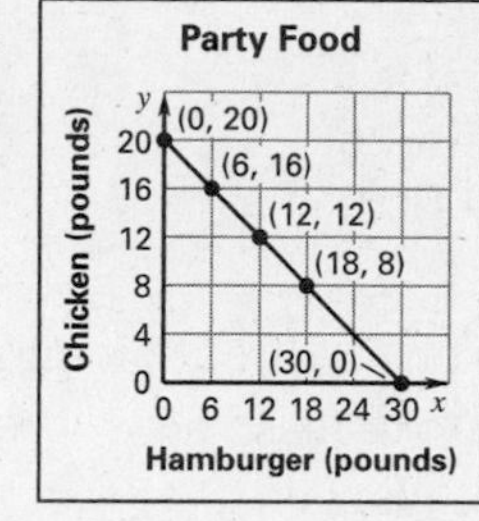

32. 20, 16, 12, 8, 0
33. $4x + 6y = 48$ **34.** $y = -\frac{2}{3}x + 8$
35.

36. 8, 6, 4, 2, 0

Practice C

1. $-3x + y = -9$ **2.** $6x - y = 12$
3. $-4x + 5y = -40$ **4.** $8x + 7y = -95$
5. $-67x - 96y = -42$ **6.** $70x + 35y = 68$
7. $x = -64$ **8.** $7y = -3$ **9.** $2x + 5y = 8$
10. $-12x + 75y = 5$ **11.** $-3x + 4y = -80$
12. $6x + 7y = -108$ **13.** $5x - y = 26$
14. $3x + y = -17$ **15.** $6x + y = 3$
16. $x + 6y = 22$ **17.** $-x + 3y = -17$
18. $x + 2y = 5$ **19.** $-2x + 3y = -27$
20. $4x + 7y = 21$ **21.** $-7x + 9y = -38$
22. $-5x + 4y = -5$ **23.** $-6x - 21y = 63$
24. $20x + 32y = -56$ **25.** $y = \frac{1}{2}, x = 0$
26. $y = 4, x = -7$ **27.** $y = -10, x = -8$
28. $y = -3, x = 5$ **29.** $1.5x + 2.5y = 60$
30. $y = -0.6x + 24$

Lesson 5.6 *continued*

31.

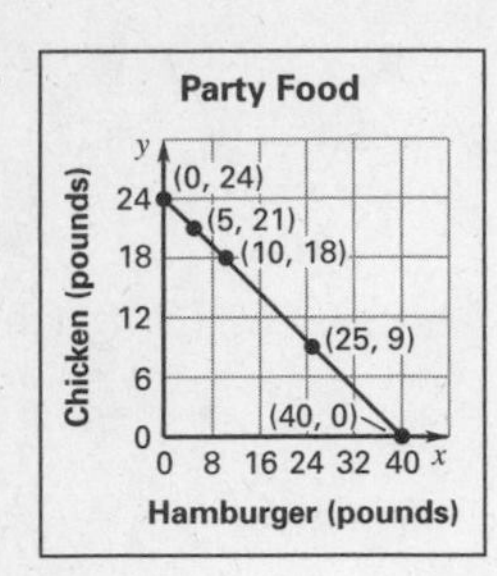

32. 24, 21, 18, 9, 0

33. (40, 0); You purchase 40 pounds of hamburger and no chicken. (0, 24); You purchase no hamburger and 24 pounds of chicken.

34. $3.75x + 5y = 75$ **35.** $y = -0.75x + 15$

36.

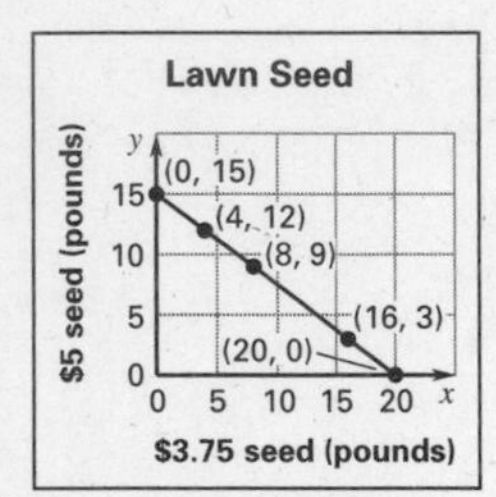

37. 15, 12, 9, 3, 0

38. (20, 0); You purchase 20 pounds of $3.75 seed and no $5 seed. (15, 0); You purchase 15 pounds of $5 seed and no $3.75 seed.

Reteaching with Practice

1. $2x - 3y = 21$ **2.** $2x - y = -8$

3. $x + 4y = 24$ **4.** $2x + y = 6$

5. $3x - y = -10$ **6.** $x + y = 3$

7.

Peaches (lb), x	0	2	4	8
Blueberries (lb), y	3	2.25	1.5	0

Real-Life Application

1. $6x + 5.5y = 200$

2.

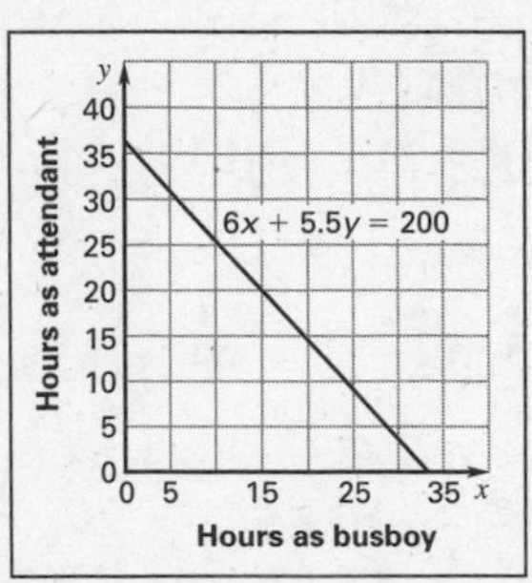

3. $1440

4. $3b + 2c = 300$ **5.**

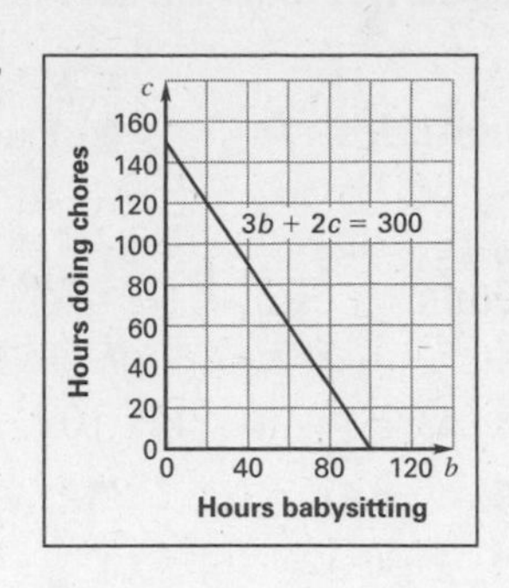

Challenge: Skills and Applications

1–4. Accept equivalent equations in standard form. **1.** $74x - 12y = 45$ **2.** $5x - 4y = -24$

3. $(p + 4)x + 3y = 2p - 4$

4. $(q + 2)x + 2y = 3q + 2$ **5.** $2x + 3y = 12$

6. 2.8 ml **7.** yes; 2.4 ml

8. $4.5x + 3.75y = 24$; 2 that are 4.5 ft and 4 that are 3.75 ft **9.** no; if he buys two 12-foot lengths, he can cut 2 shelves that are 3.75 ft and 1 shelf that is 4.5 ft from each.

Lesson 5.7

Warm-Up Exercises

1. $y = 1054x - 7886$ **2.** $y = -393x + 8995$

3. relatively no correlation

4. positive correlation

Daily Homework Quiz

1. *Sample answer:* $-7x + 10y = -15$

2. *Sample answer:* $2x + 3y = 7$

3. *Sample answer:* $20x + 25y = 500$

Lesson Opener

Allow 10 minutes.

1. Yes; most of the data points fall close to the line. **2.** No; most of the data points do not fall along or close to the line. **3.** No; most of the data points do not fall along or close to the line. **4.** Yes; most of the data points fall close to the line. **5.** Yes; most of the data points fall close to the line. **6.** No; most of the data points do not fall along or close to the line.

Practice A

1. no **2.** yes **3.** yes **4.** no **5.** yes **6.** yes

Answers

7. Linear interpolation is a method of estimating the coordinates of a point that lies between two given data points. Linear extrapolation is a method of estimating the coordinates of a point that lies to the right or to the left of all the given data points.

8. First draw a scatter plot to see if the data look approximately linear. **9.** \$346 **10.** \$786

11. about \$5.50 **12.** about \$7.00

13.

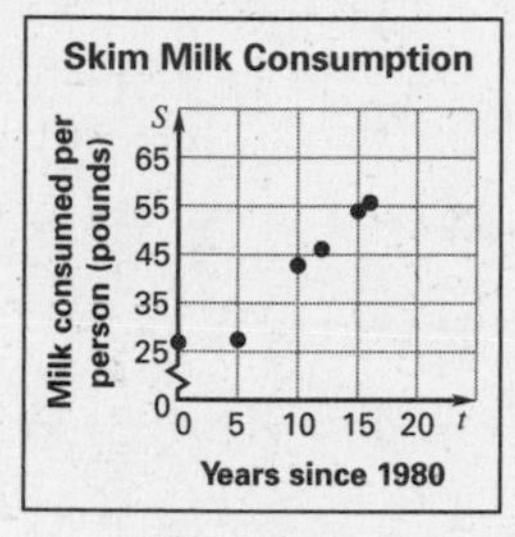

14–16. *Sample answers are given.*

14. $S = 2t + 23$ **15.** 51 **16.** 61

Practice B

1. yes **2.** no **3.** yes **4.** no **5.** yes **6.** no

7. about \$28,100 **8.** about \$61,700

9.

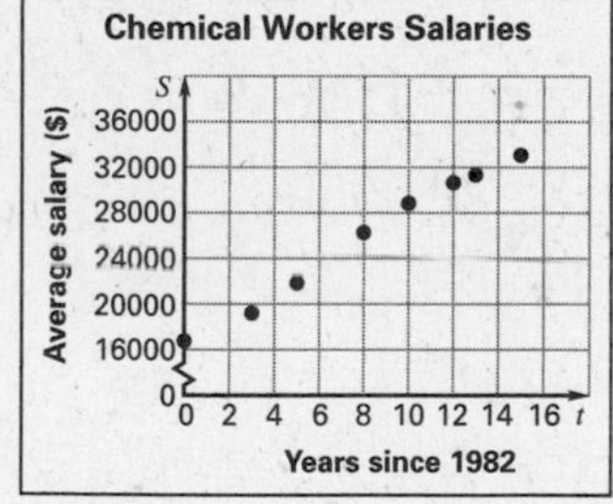

10–12. *Sample answers are given.*

10. $S = 1150t + 16,500$ **11.** about \$29,150

12. about \$36,050 **13.**

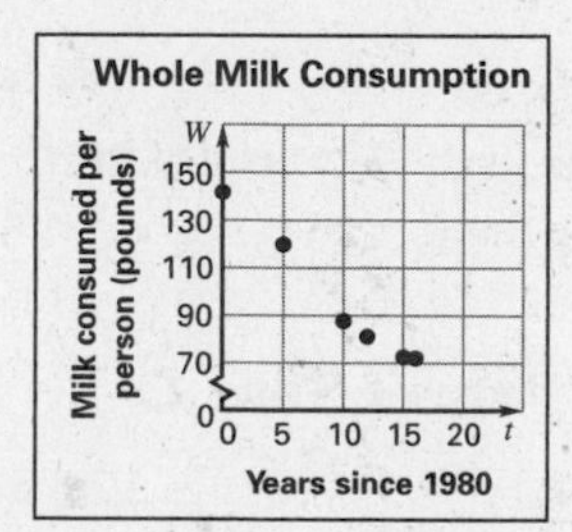

14–16. *Sample answers are given.*

14. $W = -4.6t + 140$ **15.** 75.6 **16.** 52.6

Practice C

1. yes **2.** no **3.** yes **4.** no **5.** no **6.** yes

7. \$118,770,000,000 **8.** \$62,220,000,000

9. No; Media and methods of advertising have changed considerably since 1930.

10.

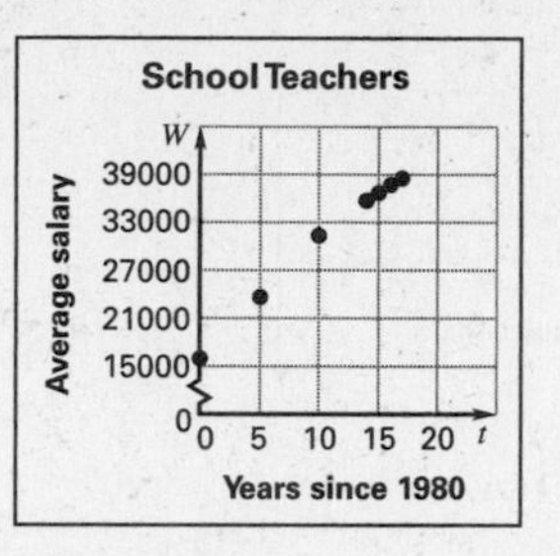

11–13. *Sample answers are given.*

11. $S = 1300t + 16,700$ **12.** about \$32,300

13. about \$41,400 **14.**

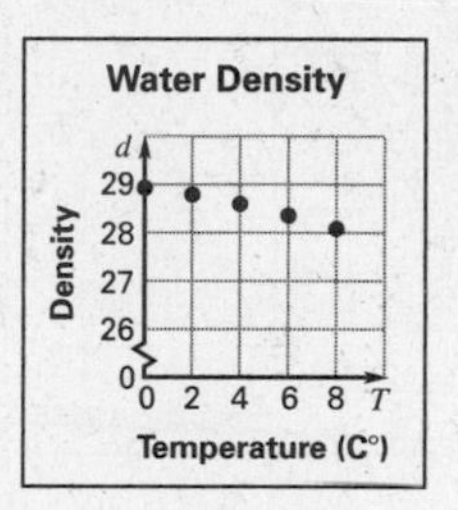

15–17. *Sample answers are given.*

15. $d = -0.14T + 29$ **16.** 28.02 **17.** 27.74

Reteaching with Practice

1. not reasonable to be represented by a linear model **2.** reasonable to be represented by a linear model

3. a.

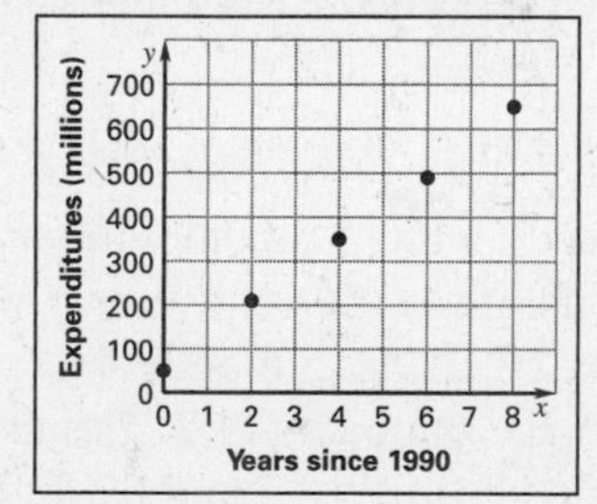

b. *Sample answer:* $y = 75x + 50$

4. *Sample answer:* \$125 million

Lesson 5.7 *continued*

Real-Life Application

1.

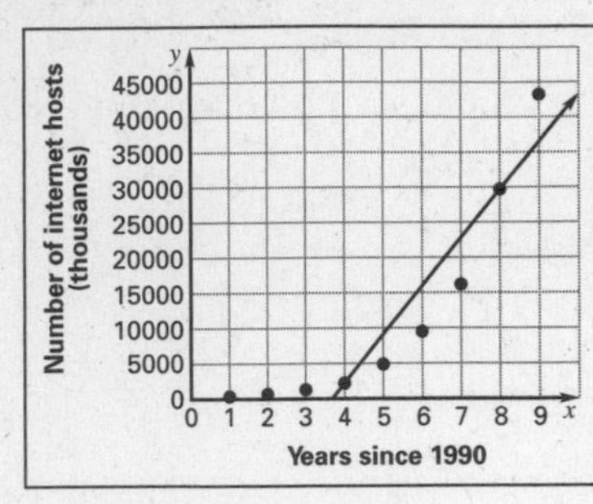

2. $y = 6863x - 25{,}235$

3. 57,000

4.

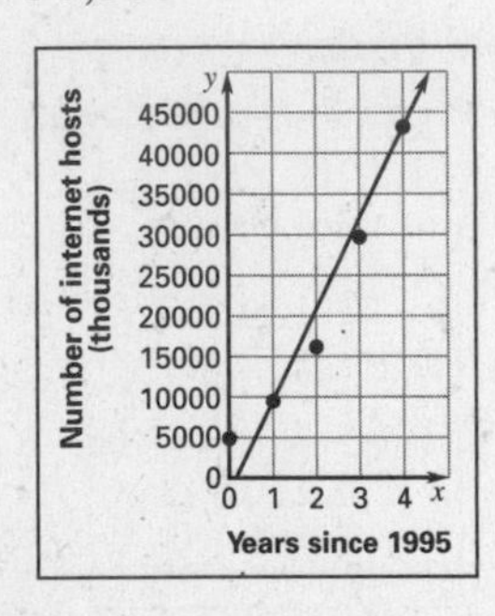

5. $y = 11{,}252x - 1780$ **6.** 77,000

Challenge: Skills and Applications

1. $y = 3x - 4$ **2.** $y = -2.1x + 7.8$

3. *Sample answer*: $y = -10.6t + 358$ (using points (0, 358) and (25, 94))

4. The exchange rate is decreasing; the yen is getting stronger.

5. $y = -10.4t + 356.7$ **6.** $y = -10.1t + 352.7$

7. Estimates based on Ex. 3 model will vary. Using sample model given above, estimate would be 40; Estimate based on Ex. 5 model: 44.7; Estimate based on Ex. 6 model: 49.7; The estimates using the models in Exs. 5 and 6 are fairly close. The estimate using the model in Ex. 3 will vary, but it is likely to be fairly close to at least one of the other models.

Review and Assessment

Review Games and Activity

1. $6x + 5y = -10$ **2.** $y = \frac{1}{2}x + 3$

3. $3x - 7y = 49$ **4.** $8x - 7y = 70$

5. $y = -\frac{13}{6}x - 7\frac{2}{3}$ **6.** $6x - 7y = -42$

7. $y = -\frac{5}{8}x + 2\frac{5}{8}$ **8.** $2x - y = 3$

9. $y = \frac{1}{2}x - 1$ **10.** $x + 3y = 9$

POPLAR ONES

Chapter Test A

1. $y = -5x + 7$ **2.** $y = 10x - 3$

3. $y = x + 1$ **4.** $y = 5x + 5$

5. $y = 70 - 7x$ **6.** $y = -2x + 6$

7. $y = 2x$ **8.** $y = \frac{2}{3}x + \frac{7}{3}$ **9.** $y = \frac{3}{2}x - 2$

10. $y = x - 5$ **11.** $y = 2x + 9$

12. $y = -\frac{3}{5}x - \frac{2}{5}$ **13.** $y = \frac{6}{5}x + \frac{7}{5}$

14. $y = -\frac{1}{2}x + \frac{11}{2}$ **15.** $y + 4 = \frac{4}{3}(x + 3)$ or $y - 4 = \frac{4}{3}(x - 3)$ **16.** $y + 4 = -\frac{1}{12}(x + 5)$ or $y + 5 = -\frac{1}{12}(x - 7)$ **17.** $5x - y = -6$

18. $3x + y = 9$ **19.** $x = 3, y = 3$

20. $x = -1, y = 2$ **21.** No **22.** Yes

Chapter Test B

1. $y = -3x + 5$ **2.** $y = 4x$ **3.** $y = \frac{3}{4}x + 3$

4. $y = \frac{2}{3}x - 2$ **5.** $y = 4 + 3x$ **6.** $y = \frac{1}{2}x + \frac{1}{2}$

7. $y = \frac{1}{2}x + \frac{7}{2}$ **8.** $y = \frac{2}{7}x + \frac{13}{7}$

9. $y = -\frac{6}{5}x + \frac{8}{5}$ **10.** $y = -3x + 9$

11. $y = \frac{1}{2}x + \frac{1}{2}$ **12.** $y = -\frac{8}{9}x - \frac{13}{9}$

13. $y = x - \frac{1}{2}$ **14.** $y = \frac{1}{3}x + \frac{5}{3}$

15. $y + 6 = \frac{1}{4}(x - 5)$ or $y + 7 = \frac{1}{4}(x - 1)$

16. $y + 3 = -\frac{12}{7}(x - 6)$ or $y - 9 = -\frac{12}{7}(x + 1)$

17. $2x - 8y = 3$ **18.** $x + 3y = -15$

19. $x = 2, y = 4$ **20.** $x = -5, y = 4$

21. $10x + 15y = 55$ **22.** Yes **23.** No

Chapter Test C

1. $y = -\frac{4}{3}x - 2$ **2.** $y = -5$

3. $y = 150 - 55x$ **4.** $y = \frac{2}{3}x - \frac{23}{3}$

5. $y = -\frac{3}{4}x$ **6.** $y = \frac{1}{3}x - \frac{11}{3}$

Review and Assessment *continued*

7. $y = -\frac{8}{7}x - \frac{11}{7}$ **8.** $y = -4x + 17$

9. $y = -\frac{2}{3}x + \frac{5}{3}$ **10.** $F = 1.8C + 32$

11. $y = -\frac{8}{9}x - \frac{13}{3}$ **12.** $y = -\frac{13}{20}x + \frac{107}{30}$

13. $y = \frac{4}{3}x + \frac{19}{3}$ **14.** $y - 5 = -\frac{4}{3}(x + 2)$ or $y + 3 = -\frac{4}{3}(x - 4)$ **15.** $y + 1 = -6\left(x - \frac{1}{2}\right)$ or $y - 6 = -6\left(x + \frac{2}{3}\right)$ **16.** $5x - 10y = 6$

17. $24x + 9y = 4$ **18.** $x = -5, y = -7$

19. $x = 6, y = -10$ **20.** $2.39x + 1.89y = 11$

21.

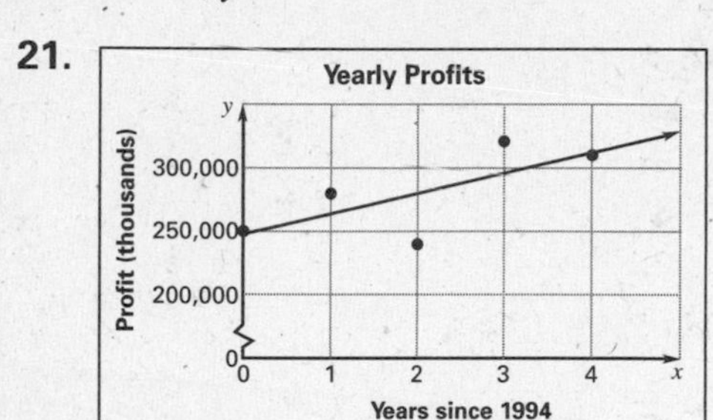

22. Answers may vary.
One possible solution: $y = 16{,}000x + 248{,}000$

23. Answers may vary.
One possible solution: $376,000,000

SAT/ACT Chapter Test

1. B **2.** C **3.** D **4.** C **5.** B **6.** D **7.** A **8.** C **9.** A

Alternative Assessment

1. a–d. Complete answers should address these points. • Explain that best-fit lines are important as they can be used to predict future data points and they help to analyze the trends of data.

a. Check graphs. **b.** • Find an equation of the best-fit line, for example, $y = 1.97x + 2.24$. For a more accurate approach students should use a graphing calculator to find the regression equation. • Explain a method (either selecting points or using a calculator). Explain that a linear model can be used to estimate data points that are not given. **c.** • Explain data is increasing over time. • Create a scenario to explain data trends. *Sample answer:* A company's sales from 1990 to 1995, in millions of dollars. Each year the company is growing, and sales are increasing. **d.** • Choose an additional x-value to substitute into the best-fit line equation to determine a future y-value. *Sample answer:* In 1996, $x = 6$, the company would make about $14 million. • Explain whether the answer is reasonable based on their specific data.

2. a. 75; -7.5; Jeremy deposits $75 weekly; Zachary withdraws $7.50 weekly. **b.** Jeremy: $y = 75x + 275$; Zachary: $y = -7.50x + 522.50$ **c.** Check graphs. **d.** Jeremy: $1775; Zachary: $372.50 **e.** 70 weeks; No; He can withdraw $5 in the last week.

3. ***Critical Thinking Solution*** They would be the same after 3 weeks. *Sample answer:* One method would be to set the two equations equal to each other, $75x + 275 = -7.50x + 522.50$, and solve for x. Another method would be to sketch the graph of both equations and find out where the two equations have the same x-value. This would occur where the two equations intersect. A third method would be to use a table of values, either on the calculator or by hand, and calculate the balances for several weeks to determine the solution.

Project: What We Read

1. books and maps **2.** *Sample answer:* (using 1990 and 1994) $y = 0.775t + 17.5$; (using 1992 and 1994) $y = 1.45t + 18.7$

3. *Sample answer:* books and maps: $19.825 billion; magazines, newspapers, and sheet music: $23.05 billion; interpolation **4.** estimates for magazines, newspapers, and sheet music

5. *Sample answer:* books and maps: $25.25 billion; magazines, newspapers, and sheet music: $33.2 billion; extrapolation **6.** Answers will depend on data from the year 2000.

Cumulative Review

1. -343 **2.** $\frac{13}{27}$ **3.** -0.981 **4.** -217

5. $l = 3w$ **6.** $t < \frac{1}{2}s$ **7.** $h = \frac{2a}{b}$ **8.** -3

9. 21.833 **10.** 2.258 **11.** $-\frac{49}{18}$ **12.** -7

13. 7.933 **14.** 33.767 **15.** 4

16. x-intercept $= \frac{7}{2}$, y-intercept $= -14$

17. x-intercept $= -\frac{5}{3}$, y-intercept $= \frac{4}{27}$

18. $-\frac{99}{350} \approx 0.2829$ **19.** $-\frac{1}{49}$ **20.** $y = 3x - 2$

21. $y = -5x + 4$ **22.** $y = 7x + 11$

23. $y = \frac{1}{5}x - 6$ **24.** $y = 8$

25. $y = -6.5x + 4.5$ **26.** $y = 4x - 18$

Review and Assessment *continued*

27. $y = -\frac{1}{4}x + 2$ **28.** $y = -3x + 7$

29. $y = \frac{4}{3}x - 3$ **30.** $y = -\frac{1}{5}x - \frac{3}{5}$

31. $y = -2x - 80$ **32.** $y = -3x + 1$

33. $y = -0.923x - 0.460$

34. a.

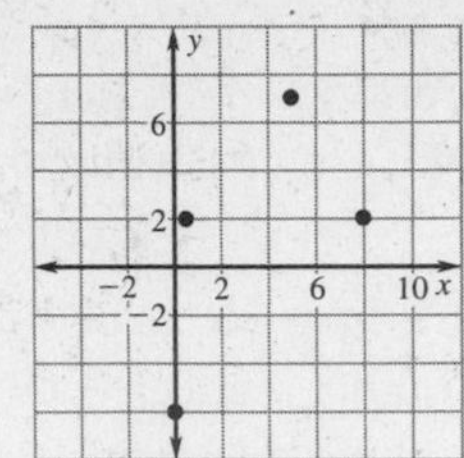

b. No correlation

35. a.

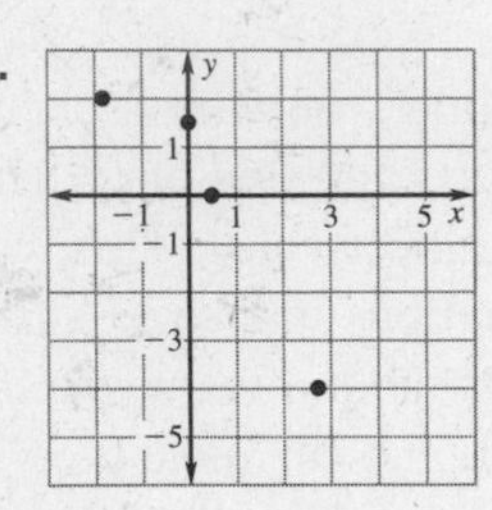

b. Negative correlation

c. *Sample Answer:* $y = -1.36x + 0.357$

36. a.

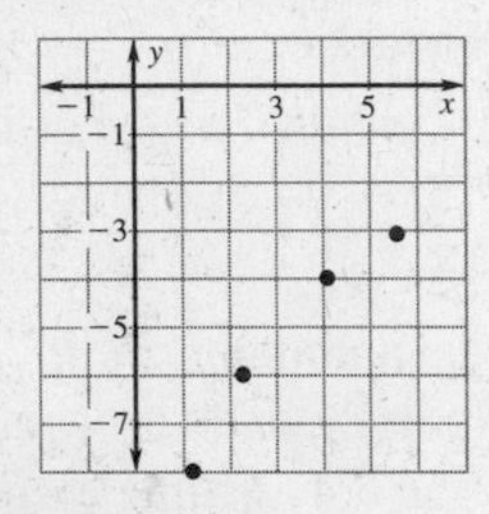

b. Positive correlation

c. *Sample Answer:* $y = 1.10x - 8.90$

37. $y - 7 = \frac{7}{2}(x - 2)$ **38.** $y - 8 = 2(x + 10)$

39. $y - 1 = \frac{2}{3}(x - 4)$ **40.** $y - 10 = -4(x + 10)$

41. $5x + y = 5$ **42.** $x + 2y = 10$

43. $12x + 2y = -9$ **44.** $4x - y = -20$

45. 8.8 million people

46.

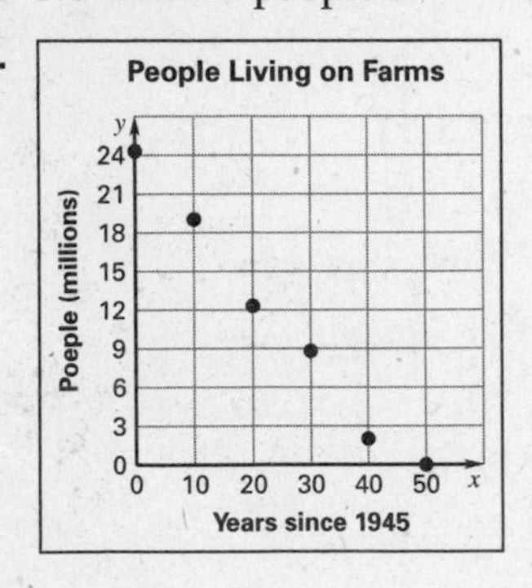

47. *Sample Answer:* $y = -0.502x + 23.6$